Advanced Antenna Technologies for Communication, IoT and Medical Applications

Edited by Albert Sabban

Published in London, United Kingdom

Advanced Antenna Technologies for Communication, IoT and Medical Applications
http://dx.doi.org/10.5772/intechopen.1009703
Edited by Albert Sabban

Contributors
Albert Sabban, C. Rimmya, Dairoku Muramatsu, Felipe Gil-Castiñeira, Joaquim Bastos, Jonathan Rodriguez, Karamot Kehinde Biliaminu, Madhan M. Ganesh, Odai Hassan Raheem Al Soad, S. Piramasubramanian, Somayeh Mohammady, V. Lavanya, Wan Chen, Zaid Shakir Kadhim Al-Shammari

First published in London, United Kingdom, 2026 by IntechOpen
IntechOpen is the global imprint of INTECHOPEN LIMITED, registered in England and Wales, registration number: 11086078, 167-169 Great Portland Street, London, W1W 5PF, United Kingdom

For EU product safety concerns: IN TECH d.o.o., Prolaz Marije Krucifikse Kozulić 3, 51000 Rijeka, Croatia, info@intechopen.com or visit our website at intechopen.com.

British Library Cataloguing-in-Publication Data
A catalogue record for this book is available from the British Library

Advanced Antenna Technologies for Communication, IoT and Medical Applications
Edited by Albert Sabban
p. cm.
Print ISBN 978-1-83635-184-9
Online ISBN 978-1-83635-183-2
eBook (PDF) ISBN 978-1-83635-185-6

Meet the editor

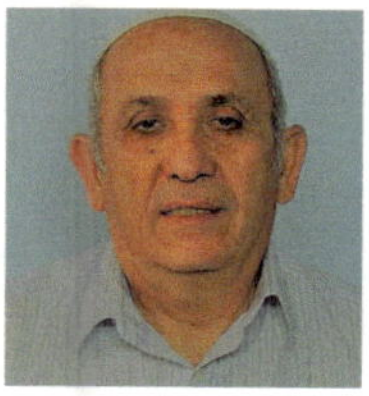

Dr. Albert Sabban holds a Ph.D. in Electrical and Computer Engineering from the Faculty of Electrical and Computer Engineering, University of Colorado at Boulder, USA (1991), and an MBA from the Faculty of Management, Haifa University, Israel (2005). He holds a BSc and MSc (Magna Cum Laude) from Tel Aviv University, Israel. He is a senior lecturer and researcher at electrical and computer engineering colleges in Israel. From 2007 to 2010, Dr. Sabban was a radio frequency and antenna specialist at biomedical high-tech companies, where he designed compact wearable medical systems. In 1976, he joined Rafael Advanced Defense Systems Ltd., Israel, where he worked as a senior researcher, group leader, and project leader in the electromagnetics department until 2007. He has published more than 100 research papers and holds several patents. He has written and edited more than 15 books and book chapters on wearable systems, antennas, and green electronics technologies for communication and medical systems. He is a senior consultant and researcher in communication systems, antennas, and green technologies for high-tech companies.

Contents

Preface

Communication and medical systems demand higher efficiency, wideband data rates, and more channel capacity. The development of advanced antenna technologies and wideband RF technologies may achieve these specifications. Advanced antenna technologies have attracted significant attention in the last decade. Advanced antenna technologies' main topics span a wide spectrum, including metamaterials, fractal technologies, active antennas, MIMO antennas, green antennas, and several types of printed antennas. The major features of advanced antennas should be wideband operation, high efficiency, green (low power consumption), and cost-effectiveness. Lightweight, small, wideband RF devices and antennas are essential for evaluating advanced communication devices, RF systems, IoT devices, and direction-finding modules. Communication links suffer from fading effects, frequency interference, and multipath losses. The book presents advanced wearable antennas, printed antennas, and metamaterial antennas. Green technologies and energy-harvesting techniques are also presented in the book.

The mobile communication, 5G, RFID, and IoT industries have been in continuous growth over the last decade. There is a continuous increase in demand for wearable sensors, RFID tags, and devices. Wearable technology has several applications in personal wireless communication, IoT, RFID, sports, and medical devices, as presented in [1–7]. Medical devices and systems were developed to monitor patient health, as presented in several books and papers [1–8]. Wearable technology provides a powerful new tool for medical services, surgical rehabilitation services, and IoT systems. Wireless Body Area Networks, WBANs, can record electrocardiograms, measure body temperature and blood pressure, measure heartbeat rate, arterial blood pressure, electro-dermal activity, and other healthcare parameters. Wearable devices will be an important part of individuals' daily lives in the next decade.

Advanced antennas are used in modern wearable devices, IoT devices, and medical wearable devices. Medical sensors, RFID tags, and body-worn sensors are called wearable devices. Wearable technology is a multidisciplinary developing field. Knowledge in bioengineering, electrical engineering, software engineering, and mechanical engineering is needed to design and develop wearable communication and medical systems. Wearable medical systems and sensors measure and monitor physiological parameters of the human body. Biomedical systems in the vicinity of the human body may be wired or wireless. Many physiological parameters may be analyzed by using wearable medical systems and sensors. Wearable medical systems and sensors can measure body temperature, heart rate, blood pressure, sweat rate, and other physiological parameters of the person wearing the device. Wearable technology may offer scanning and sensing features that mobile phones and laptop computers do not. Wearable technology usually has communication capabilities, allowing users to access information in real time. Several wireless technologies are used to handle data collection and processing by medical systems. The collected data may be stored or transmitted to a medical center for analysis. Wearable devices gather raw data, which is fed to a database or to a software application for analysis. This analysis may result in a response that alerts a physician to contact a patient experiencing abnormal symptoms. A message may be sent to a person who achieves a fitness goal.

Examples of wearable devices include belts, headbands, smart wristbands, watches, glasses, contact lenses, e-textiles and smart fabrics, jewelry, bracelets, and hearing aid devices.

Usually, wearable communication devices consist of a wearable antenna, a transmitting unit, a receiving unit, and a data processing unit. Wearable technology may influence the fields of transportation, health, and medicine, IOT, fitness, aging, disabilities, education, finance, gaming, entertainment, and music.

Figure 1 presents a medical monitoring health system with wearable antennas. The proposed antennas may be placed on the patient's body as shown in **Figure 1**.

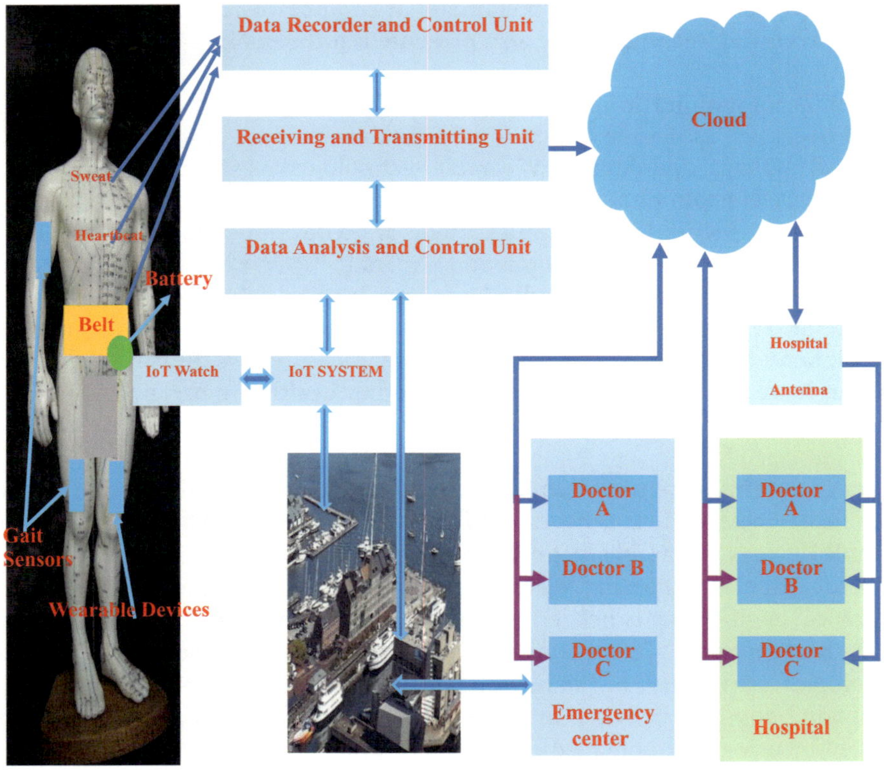

Figure 1.
A medical monitoring health system with wearable antennas.

Wearable technology provides a powerful new tool for medical services, healthcare monitoring systems, surgical rehabilitation services, sports, and IoT systems. Wireless Body Area Networks, WBANs, can record electrocardiograms, measure body temperature and blood pressure, measure heartbeat rate, arterial blood pressure, electro-dermal activity, and other healthcare parameters.

Reference

[1] A. Sabban (2015). "Low visibility Antennas for communication systems", TAYLOR & FRANCIS group, 2015, USA

[2] Albert Sabban (2016)."Wideband RF Technologies and Antenna in Microwave Frequencies", Wiley Sons, July 2016, USA

[3] A. Sabban, "New Wideband printed Antennas for Medical Applications" I.E.E.E Journal, Trans. on Antennas and Propagation, Vol. 61, No. 1, pp. 84-91, January 2013

[4] D. G. Fang, Antenna Theory and Microstrip Antennas. CRC Press, FL, USA, September 2019. ISBN 9780367384678

[5] A. Sabban, "Small Wearable Meta Materials Antennas for Medical Systems", The Applied Computational Electromagnetics Society Journal, Vol. 31, No. 4, April 2016

[6] A. Sabban, "Wearable Circular Polarized Antennas for Health Care, 5G, Energy Harvesting, and IoT Systems", Electronics 2022, 11(3),427; https://doi.org/10.3390/electronics11030427 , 30 Jan 2022

[7] Sabban, A. Green Wearable Sensors and Antennas for Bio-Medicine, Green Internet of Things, Energy Harvesting, and Communication Systems. Sensors 2024, 24, 5459, https://doi.org/10.3390/s24175459

[8] Jamil, F.; Ahmad, S.; Iqbal, N.; Kim, D.-H. Towards a Remote Monitoring of Patient Vital Signs Based on IoT-Based Blockchain Integrity Management Platforms in Smart Hospitals. Sensors 2020, 20, 2195, https://doi.org/10.3390/s20082195

Dr. Albert Sabban
Engineering,
Braude College,
Karmiel, Israel

Section 1

Advanced Antenna Technologies

Chapter 1

Advanced Green Wearable Meta Materials and Fractal Antennas for Communication, IoT and Healthcare Systems

Albert Sabban

Abstract

Communication and healthcare systems demand higher efficiency, wideband data rates, and more channel capacity. These specifications may be achieved through the development of advanced antenna technologies and wideband RF technologies. Advanced antenna technologies have attracted significant attention in the last decade. The main topics of advanced antenna technologies present a wide spectrum of research areas, such as metamaterials, fractal, active antennas, MIMO antennas, green antennas, and several types of printed antennas. The major features of advanced antennas should include wideband, high efficiency, green antennas with low power consumption, and affordability. Lightweight, small wideband RF devices and antennas are significant in evaluating advanced communication devices, RF systems, Internet of Things (IoT) devices, and direction-finding modules. Advanced wearable slot antennas, printed antennas, and energy harvesting techniques are presented in this chapter. Fractal and metamaterials can be employed to improve the electrical performance of small sensors and antennas. By employing green materials in the design of communication systems and connecting an energy harvesting unit to the antenna, a green communication system may be developed. A metamaterial antenna with CSSR and metallic strips is presented in this chapter. The antenna bandwidth for VSWR better than 2.3:1 is 48%, and the antenna gain is around 7dBi. The metamaterial antenna's resonant frequency on the human body is shifted by 2–4% in different locations on the human body.

Keywords: wearable antennas, metamaterial antennas, fractal antennas, energy harvesting, IoT, medical systems

1. Introduction

Small, printed antennas are crucial in the development of wearable sensors, 5G, and medical devices. Low efficiency and narrow bandwidth are the major disadvantages of printed small antennas, as presented in [1–4]. Printed small fractal and metamaterial sensors are employed in wearable Internet of Things (IoT), 5G, and medical systems and were analyzed, developed, and discussed in books and journals (see [2–6]).

Compact, efficient, wideband wearable antennas, such as printed dipoles, loops, slots, patches, and FIPA antennas, are presented in [2–5]. Fractal and metamaterials can be employed to improve the electrical performance of small sensors and antennas. Metamaterials consist of artificial materials and periodic resonant structures that characterize the electrical and mechanical performance of the metamaterial. By employing green materials in the design of communication systems and connecting an energy harvesting unit to the antenna, a green communication system may be developed. In [7–19], periodic structures and metallic posts, such as periodic circular split-ring resonators (CSRRs), are used to produce materials with a required dielectric constant value and permeability. Metamaterial technology can be used to evaluate compact, wideband, low-cost wearable antennas for wireless RF devices, 5G and 6G, medical sensors, healthcare monitoring sensors, and other wearable devices [12–16]. The gain and bandwidth of the metamaterial microstrip antenna presented in [6] are similar to those of conventional patches. In [8], metamaterial structures with negative permittivity are developed and presented. In [9], the authors analyze and measure the polarity of SRR structures. A transmission-line antenna with metamaterial segments is discussed in [12]. Compact metamaterial patches and printed antennas for smart devices, wearable sensors, and medical devices are presented in [15–23]. In [2, 3, 18] measurement setups and measured results of antennas in proximity to the user's body are discussed and given. Fractal antenna theory and applications are discussed in books, journals, and published research reports [20, 24–28]. Fractal antennas employ similar fractal structures to maximize the antenna's effective area. Usually, fractal antennas have small volume and are compact. Fractal antennas may have a wide band and can be used in smart cellular devices and RF systems. Several wearable and fractal antennas and sensors are analyzed and presented in [25–35]. Small antennas are analyzed and presented in [36]. Healthcare sensors track patients' medical status, such as heartbeat, sweat, and body temperature. A hospital IoT-based management platform used to monitor patients' health is presented in [37]. Wearable medical devices may monitor and test a patient's daily health [37–42]. In [43], dual-polarized wearable radiators for wearable healthcare devices and RF systems are discussed. Wearable active wideband self-powered sensors and antennas for smart grid (SG), communication, and healthcare applications are discussed and presented in [21–23, 38–43] and [44–53]. Challenges and future trends in wireless power transfer for biomedical implants are presented in [54]. A review of reconfigurable antenna types and mechanisms for wireless communications: IoT applications is presented in [55]. Implantable antenna designs for healthcare are presented in [56, 57].

1.1 Green materials used in electronic, RF, antenna systems, and medical applications

Green materials should be environmentally friendly. The most popular and important green materials are aluminum, iron alloy, borosilicate glass, graphene, and biomaterials.

1.1.1 Aluminum advantages as green materials

1.1.1.1 Low weight

Aluminum density is around 2.6–2.8 g per cm^3 and it is called a light metal. Aluminum density is approximately one-third of the density of steel.

1.1.1.2 Durable strength

Aluminum alloys are employed in several cases where strength and durability are needed, such as in the engineering of airplanes, satellites, and rockets. Aluminum's strength is increased by adding silicon, magnesium, and lithium to pure aluminum. The new compound is a stronger metal, even stronger than steel.

1.1.1.3 Resistance from corrosion

Aluminum forms a thin but effective oxide layer that prevents further oxidation. Aluminum corrosion resistance is good in most environments when untreated. Aluminum's oxide layer repairs itself immediately if damaged mechanically.

1.1.1.4 Conductive

Pure aluminum has a thermal conductivity of about 235 watts per kelvin per meter and an electrical conductivity of about 38 million siemens per meter at 25°C.

1.1.1.5 Recyclable

Recycling one tonne of aluminum saves around 10 tonnes of CO_2 emissions in a quick process. Aluminum can be recycled several times; it can be melted down multiple times and reformed without losing any quality. Recycled aluminum can be used again in less than a few weeks.

1.1.1.6 Affordability

Compared to other metals, the price of aluminum is affordable for production and has many technical advantages.

1.1.2 Borosilicate glass advantages as green materials

Borosilicate glass is a strong glass and can withstand high temperatures. Borosilicate is a type of glass made with silica and boron trioxide. Borosilicate glass is known by various names, including Pyrex, Duran, Borosil, Glassco, Supertek, Suprax, Simax, Bellco, and others. It is resistant to thermal shock and can be used in several applications, such as electronics and labware. Borosilicate glass was invented by a German glass maker the end of the nineteenth century. Borosilicate glass is created by combining and melting 13% boric oxide, 80% silica sand, 4% soda ash, and 3% alumina. It has a low coefficient of thermal expansion, approximately 3×10^{-6} K^{-1} at 20°C, and melts at about 1,650°C. Applications include implantable drug pumps, cochlear implants, veterinary tracking devices, neurostimulators for the treatment of epilepsy, and physiological sensors. Borosilicate glass is also used in the production of high-power vacuum tubes and glass transmitting tubes that operate at high temperatures. Additionally, borosilicate glasses are utilized in the semiconductor industry for the fabrication of microelectromechanical systems (MEMS). Many high-quality flashlights use borosilicate glass for their lenses.

1.1.3 Recycled metals and iron alloys

1.1.3.1 Recycled metals

Recovering metals from e-waste reduces mining impact.

1.1.3.2 Examples

- Recycled gold, silver, and copper
- Recycled aluminum and steel

1.1.3.3 Advantages

- High-value material recovery
- Drastically reduced environmental impact
- Lower energy than mining

1.1.3.4 Applications

- Circuit boards
- Wiring
- Casings

Iron alloys are exceptionally durable and can be recycled easily. Iron alloys are yellow materials.

1.1.4 Graphene advantages as green materials

Graphene is a one-atom-thick crystalline form of carbon. Graphene can be considered a green material if it is produced using physical processes, with no associated chemical solvents. Graphene can be used as a green material in several electronic applications. Graphene is an eco-friendly, nonbiological material used in the electronics industry. Graphene is one of the strongest materials produced on Earth. Graphene has several characteristics, including thermal conductivity twice that of diamond, high electron mobility that is a hundred times faster than silicon, electrical conductivity much better than copper, and absorption of only 2.3% of reflected light. The intrinsic dielectric constant of the bilayer graphene interface is around 6 ± 2. Graphene specifications compared to copper and silver are listed in **Table 2**. Graphene can be used in many electronics applications, from faster transistors and bendable phones to improved touch screens and advanced circuitry for computers. Several researchers claim that the small size and unique physio-chemical properties of graphene pose potential risks to the health of animals, humans, and the environment. Based on a literature review presented in [18], graphene is a chemically persistent substance that could exert considerable toxicity.

A study presented in [19] found that micrometer-thick sheets of a few graphene layers can penetrate cell membranes at the sheets' corners and asperities, of which numerous exist along the irregular edges of fabricated graphene. This piercing of cells illustrates the inherent potential of graphene to cause harm to organisms. To minimize the risks of graphene, researchers have provided three recommendations.

1.1.5 Biodegradable polymers

Biodegradable polymers are used to replace traditional plastics in casings, circuit boards, and packaging.

1.1.5.1 Examples

- PLA (polylactic acid)
- PHA (polyhydroxyalkanoates)
- PBS (polybutylene succinate)
- Cellulose-based plastics

1.1.5.2 Advantages

- Renewable and biodegradable
- Reduced carbon footprint
- Lower toxicity

1.1.5.3 Applications

- Electronic housings
- Disposable sensors
- Flexible PCBs
- Packaging materials

1.1.6 Green printed circuit boards, PCB materials

Printed circuit boards friendly to the environment.

1.1.6.1 Examples

- Halogen-free laminates (HF-PCB)
- Bio-resins (epoxy-free)

- Natural Fiber substrates
    - Flax Fiber
    - Hemp Fiber
    - Kenaf Fiber

1.1.6.2 Advantages

- Lower toxicity
- Safer disposal
- Improved biodegradability (for natural Fibers)

1.1.6.3 Applications

- Low-power consumer electronics
- Wearables
- Disposable medical devices

1.1.6.4 Lead-free solders and pastes for PCBs

Replacements for toxic lead-based solder

1.1.6.5 Examples

- **Sn–Ag–Cu (SAC)** alloys
- **Sn–Cu**
- **Sn–Bi**

1.1.6.6 Advantages

- Eliminates lead contamination
- Complies with RoHS standards

1.1.6.7 Applications

- All modern electronic assemblies

1.1.7 Green semiconductor materials

Materials that reduce toxicity or energy use in semiconductors.

1.1.7.1 Examples

- **Silicon carbide (SiC)**
- **Gallium nitride (GaN)**
- **Organic semiconductors (pentacene, P3HT)**
- **Perovskite materials (for solar cells)**

1.1.7.2 Advantages

- High energy efficiency
- Lower fabrication energy
- Flexible and printable (organic types)

1.1.7.3 Applications

- Power electronics
- Solar panels
- OLED displays
- Printed electronics

Table 1 summarizes applications of green materials.

Category	Examples	Main benefits	Key applications
Biodegradable polymers	PLA, PHA, PBS	Renewable, biodegradable	Casings, flexible PCBs
Recycled plastics	rPET, rABS	Reduce waste	Housings, packaging
Green PCB materials	Halogen-free, natural Fiber	Low toxicity	Wearables, low-power devices
Lead-free solders	SAC, Sn–Cu	Eliminates lead	All electronics
Green semiconductors	SiC, GaN, organics	Energy-efficient	Power devices, displays
Natural fibers	Bamboo, flax	Sustainable, strong	Cases, speakers
Green battery materials	LFP, solid-state	Safer, lower toxicity	Electronics, EVs
Recycled metals	Cu, Au, Al	Reduces mining	PCBs, wiring
Eco-friendly coatings	Water-based, bio-resin	Low VOC	Protection layers

Table 1.
Green materials.

2. Green microstrip antenna with CSSR, circular split ring resonator

A single layer microstrip antenna with 15 split ring resonators has been designed by employing full-wave electromagnetic software [58]. The metamaterial sensor is etched on a substrate with a dielectric constant of 2.25 and is 0.16 cm thick. The size of the antenna presented in **Figure 1** is 36 × 20 × 1.6 cm. For S11 lower than -6 dB, the antenna bandwidth is around 9%. The measured S11 parameters on the human body are shown in **Figure 2**. The antenna beamwidth is around 74°, and the antenna directivity is eight. The patch gain is around 7.7dBi. The antenna radiation pattern is shown in **Figure 3**.

The metamaterial patch efficiency is around 96%. The theory of metamaterial antennas was discussed in [1–3]. As shown in **Figure 1,** an energy harvesting unit is connected to the antenna and consists of a rectifying circuit and a rechargeable battery [3–6]. The harvesting module and the sensor generate a green cordless wearable device. The rectifying circuit converts electromagnetic AC energy to direct current (DC energy). The antenna can be employed in green RF, medical, and IoT systems.

The antenna can be employed in green communication, medical, and IoT systems.

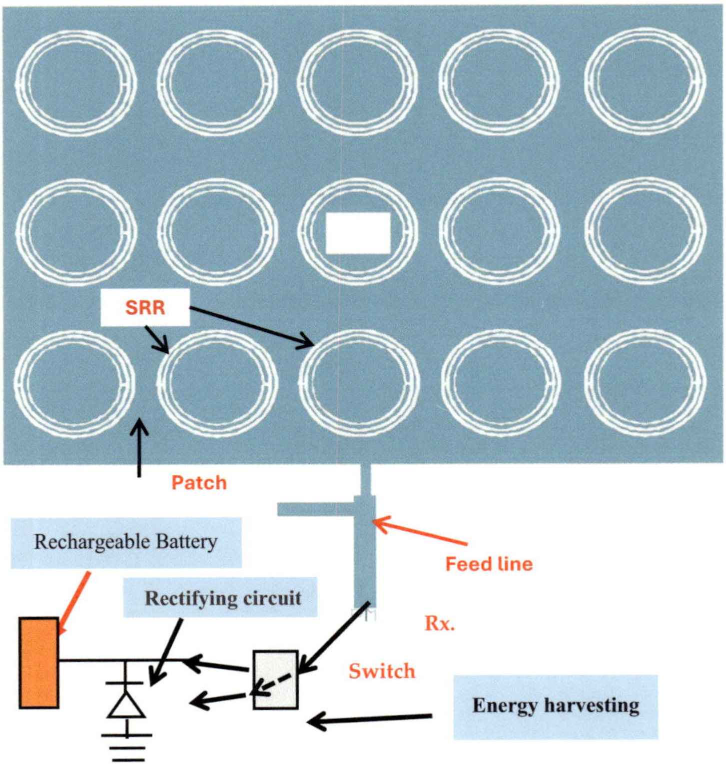

Figure 1.
Green metamaterial antenna with CSSRs [2].

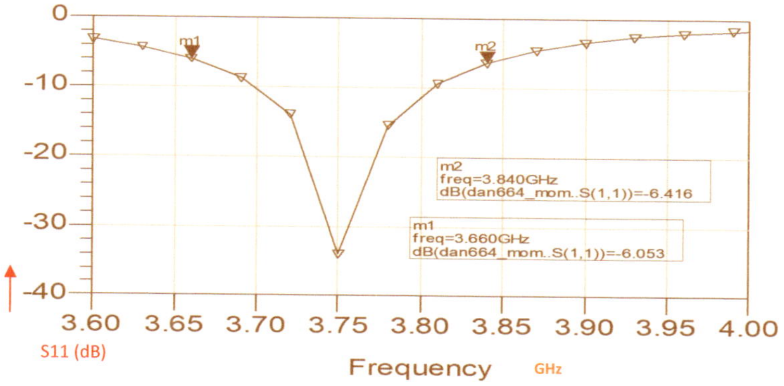

Figure 2.
Green metamaterial antenna with metallic strips, S11 results [2].

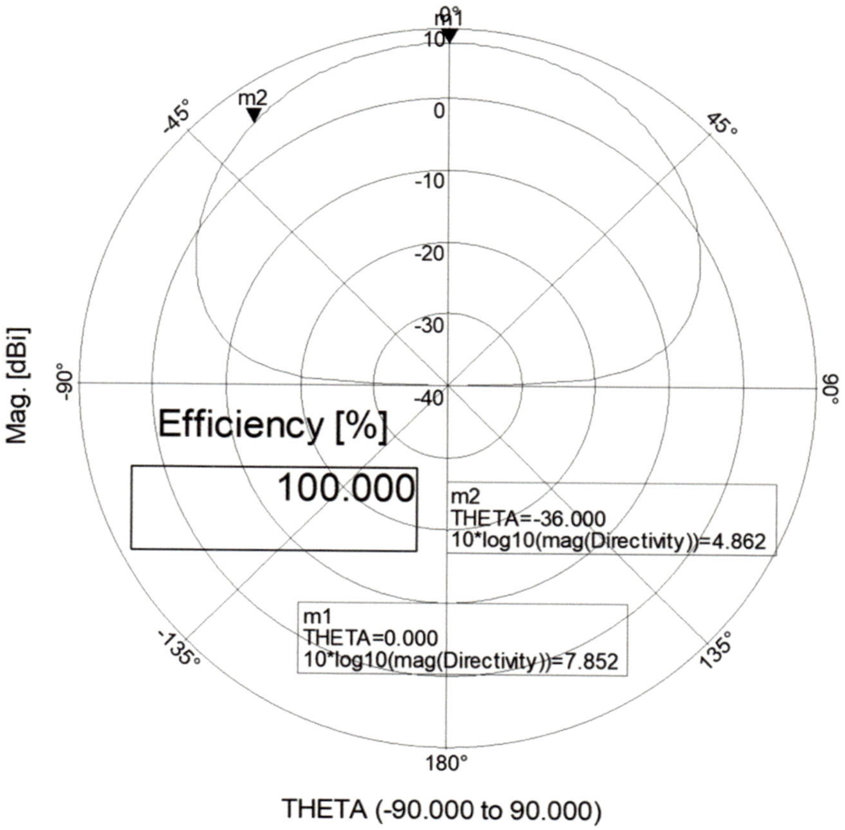

Figure 3.
Radiation pattern of the green metamaterial antenna with metallic strips [2].

3. Green wearable dual polarized metamaterial antenna

A green dual polarized metamaterial antenna with CSRRs and metallic strips is presented in **Figure 4**. The antenna is a double-layer antenna with a thickness of 3.2 mm. The antenna feed network and the metallic strips are etched on the first layer. On the second layer, a horizontally polarized dipole with CSRRs and a vertically polarized slot antenna are etched. The metamaterial antenna dimensions are 200 x 50mm. The resonant frequency of the metamaterial antenna is around 325 MHz. In **Figure 5,** the simulated S11 and the antenna gain are shown. For S11 lower than -6 dB, the measured antenna bandwidth is around 48%. The measured directivity and gain of the metamaterial antenna are 6 ± 1dBi as presented in **Figure 6**. **Figure** 7 shows the antenna gain and radiation pattern. The matching network of the metamaterial antenna was optimized to yield S11 lower than -6 dB for frequencies from 180 MHz to 400 MHz, with a 60% bandwidth, as presented in **Figure 8**.

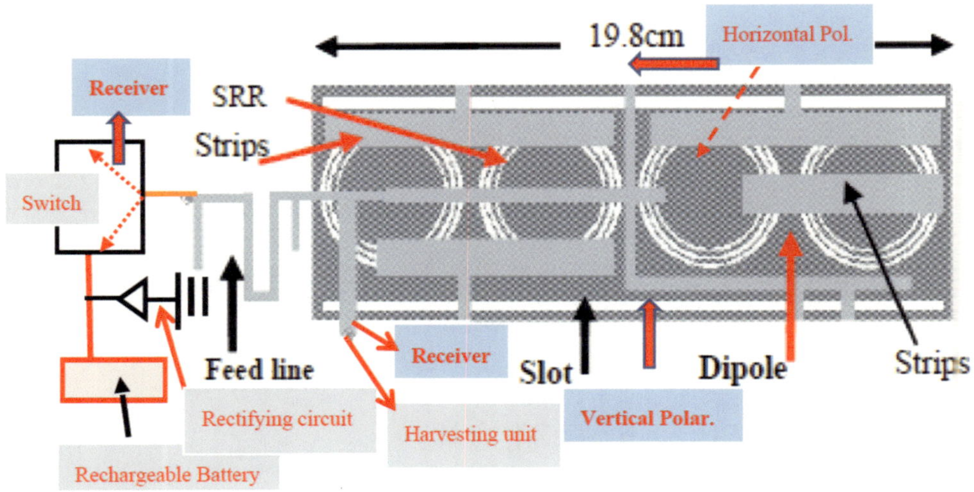

Figure 4.
Green metamaterial antenna with metallic strips [2].

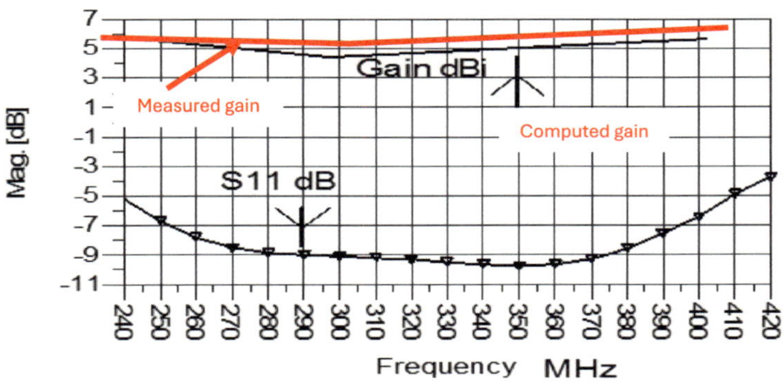

Figure 5.
Gain and S11 of the green metamaterial antenna with metallic strips [2].

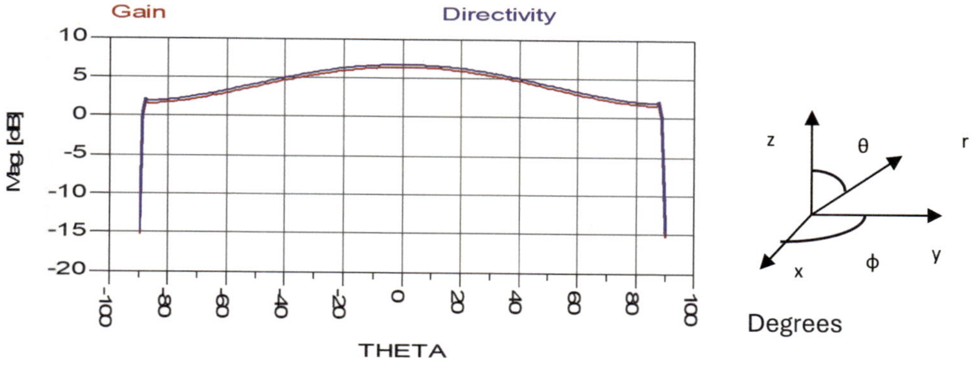

Figure 6.
Radiation pattern and gain of the metamaterial antenna with metallic strips [2].

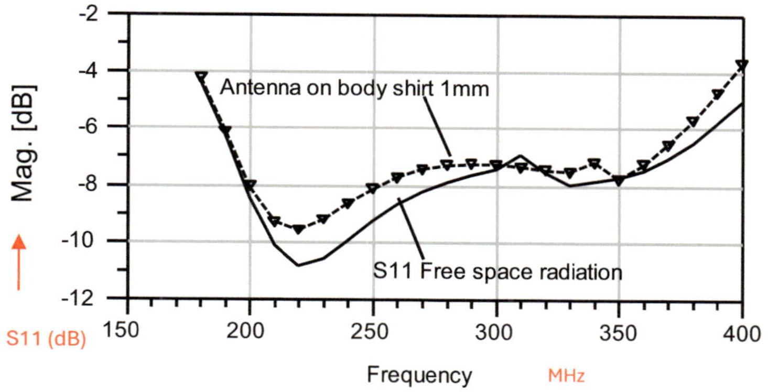

Figure 7.
S11 of the wideband-optimized antenna with metallic strips on the human body [2].

The harvesting dual-mode module and the dual-polarized antenna generate a green cordless wearable device, as shown in **Figure 4**. The rectifying circuit converts electromagnetic AC energy to direct current (DC energy). The antenna can be employed in green RF, medical, and IoT systems. The harvesting module can charge the battery when the switch is connected to the harvesting unit.

4. Compact green circular polarized metamaterial antenna

The circular metamaterial patch antenna with 18 CSRRs can be employed as a sensor in healthcare and communication systems. The CSRR strip width is 0.2 mm and the outer ring diameter is 5.2 mm, as shown in **Figure 8a**. The circular patch with 18 CSRRs, shown in **Figure 8b** and **8c**, are printed on substrate with dielectric constant of 2.2, 0.16 cm thick. The antenna diameter is 3.6 cm. The antenna resonates in the frequency range from 2.4 GHz to 2.9 GHz. Eq. (1) [1], may be used to compute TM11 mode resonant frequency. Eq. (2) may be used to compute the radius $a_e$of the circular patch, the light velocity in vacuum is c. $\varepsilon_e$is the effective dielectric

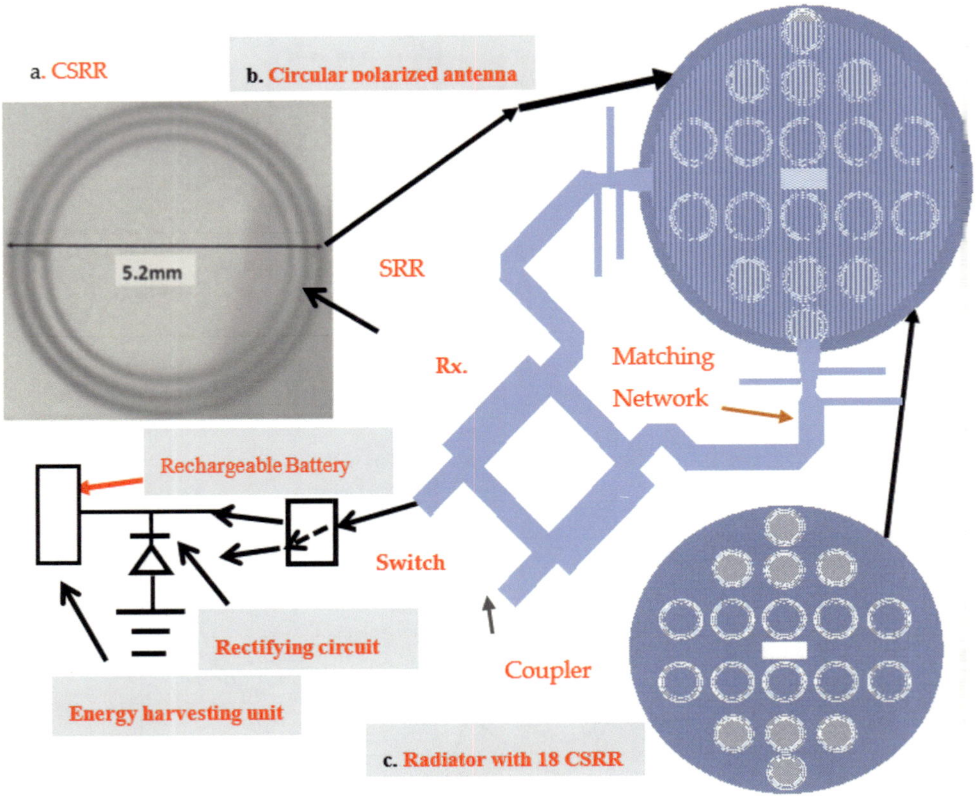

Figure 8.
(a) Photo of a CSRR. (b) Green circular polarized metamaterial antenna. (c) The radiating circular patch [58].

constant. The CSRRs design considerations was discussed in [2–19]. The radius of the patch without CSRR is higher by 24% than the radius of the patch with CSRRs. The Patch and the CSRR dimensions were optimized by using electromagnetic software [58]. The bandwidth of the circular polarized antenna is around 23%, computed and measured, for S11 lower than - 6 dB as shown in **Figure 9**. As presented in **Figure 10**, the metamaterial antenna gain is around 8 dBi with 80° beamwidth. The gain of circular patch antennas without CSRR is around 5–6 dBi. Lower by 2–3 dB than the gain of the antenna presented in **Figure 8**. The harvesting dual mode module and the circular polarized antenna generate a green cordless wearable device as shown in **Figure 8**. The rectifying circuit converts electromagnetic AC energy to direct current (DC energy). The antenna can be employed in green RF, medical, and IoT systems. The harvesting module can charge the battery when the switch is connected to the harvesting unit.

$$f = \frac{1.8412c}{2\pi a_e \sqrt{\varepsilon_e}} \tag{1}$$

$$a_e = \frac{1.8412c}{2\pi f \sqrt{\varepsilon_e}} \tag{2}$$

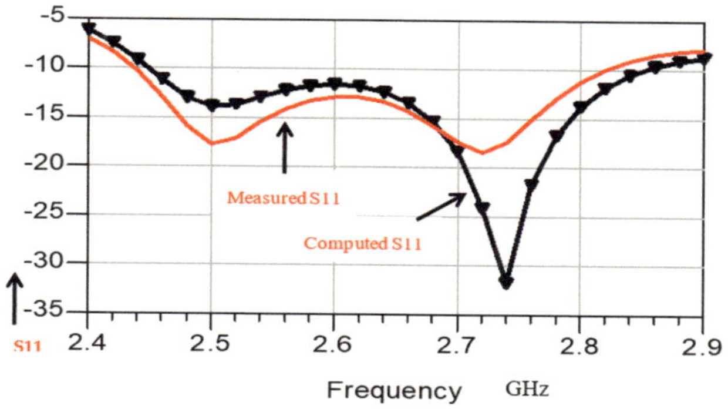

Figure 9.
Simulated and measured S11 of the circular polarized patch antenna [53].

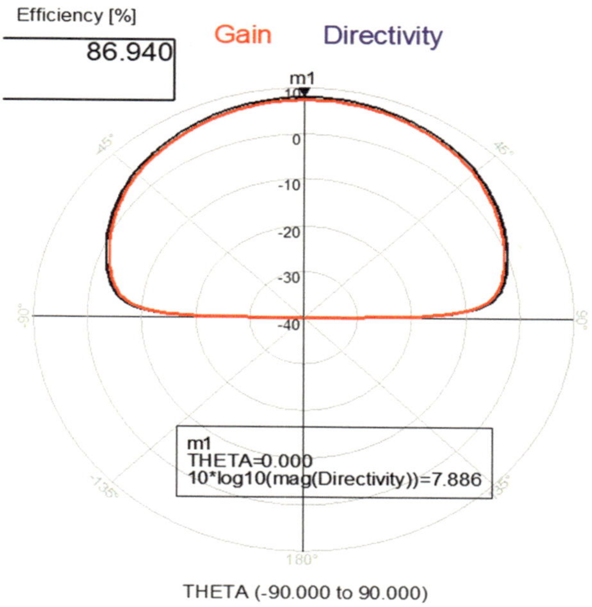

Figure 10.
Radiation pattern of the metamaterial circular polarized patch at 2.6 GHz [53].

4.1 A 3 db 90° coupler

Figure 11 shows a microstrip 3 dB 90° coupler. The dimensions of the coupler are given in **Table 2**. S11 results of the branch coupler are presented in **Figure 12**. The measured S13 transfer parameter is around −3 dB ± 0.5dB. The coupler phase difference between ports 2 and 3 is 90° ± 5°.

5. Fractal structures and antennas

Fractal structures are presented in **Figure 13**. **Figure 13a** presents a line. **Figure 13b** presents a bent line. **Figure 13c** presents a bent line fractal structure. A fractal filling structure is presented in **Figure 13**.

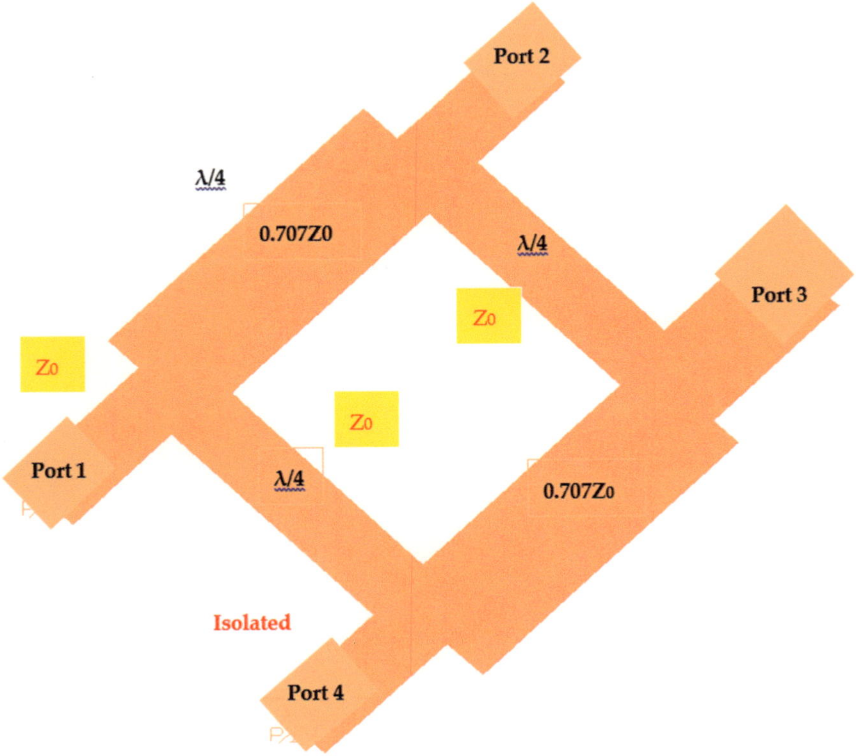

Figure 11.
*The –3 dB **90°** coupler diagram [53].*

Microstrip line impedance	Line width mm	ε_e	Length (λ/4) mm
Z0 = 50 Ω	2.95	3.45	14.85
Z0 = 35.4 Ω	5.05	3.55	14.45

Table 2.
3 db 90° coupler parameters and design

Figure 14 presents a photo of a fractal double-layer patch. **Figure 14a** presents the antenna resonator. **Figure 14b** presents the antenna radiator. The double-layered fractal antenna bandwidth for S11 lower than –6 dB is around 22%. The double-layered fractal antenna gain is 7.8 dBi.

6. Advanced wearable and implanted antenna technologies and applications

Up-to-date wearable antennas are used almost in everyday life tasks. Some examples are presented in this section. Wearable antennas are used in monitoring health, military applications, sports tech, smart textiles, and IoT applications.

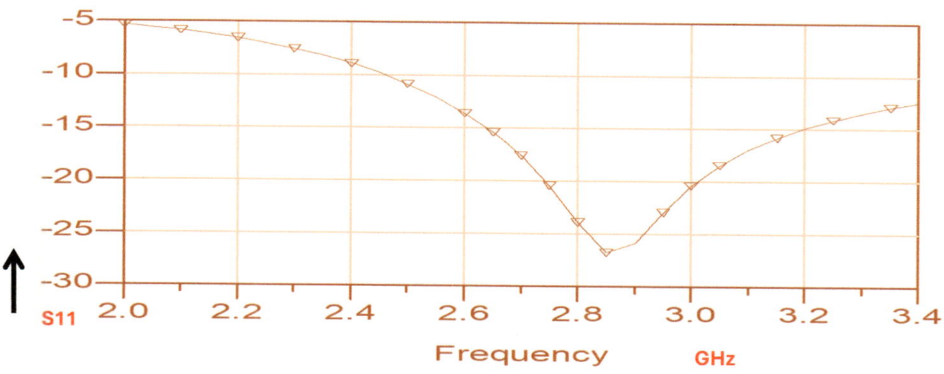

Figure 12.
The −3 dB ***90°*** *coupler S11 parameter [53].*

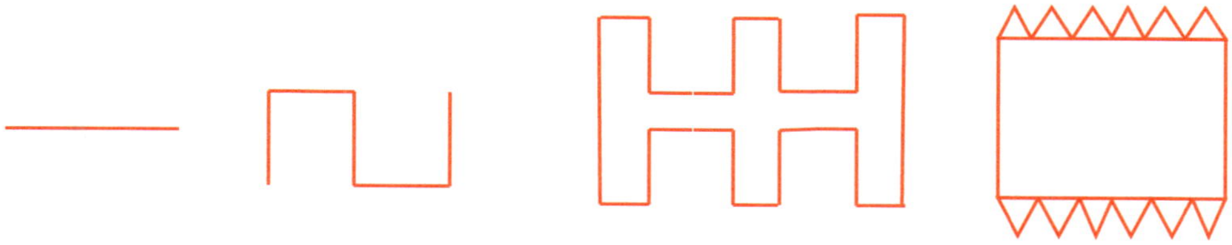

Figure 13.
Fractal structures: (a) single line, (b) motif of bent line, (c) several bent line structures, (d) fractal filling structure [2].

Figure 15 presents a healthcare monitoring network, WBAN networks, with wearable sensors and antennas.

One of the main goals of wearable medical systems is to increase disease prevention. The usage of wearable medical sensors by patients may result in better, low-cost medical treatment.

6.1 Advantages of wearable technologies

6.1.1 Healthcare, safety, and fitness monitoring

- Tracks heart rate, steps, sleep, calories, SpO_2, and several other features.
- Help users detect irregular patterns such as heartbeats, arrhythmias, and temperature changes.
- Fall detection, emergency SOS, and GPS tracking for the vulnerable.
- Wearable ECGs, glucose monitors, insulin pumps, and other digital health devices improve chronic disease management.

Figure 14.
Fractal double-layer patch. (a) Fractal resonator. (b) Patch radiator [2].

6.1.2 Real-time data and convenience

- Immediate feedback during exercise, work, or daily activities.
- Improves decision-making and lifestyle adjustments.
- Smartwatches allow calls, messages, navigation, and payments without using a phone.

6.1.3 Enhanced productivity

- Notifications, reminders, scheduling, and quick communication.

6.1.4 Personalized features and insights

- AI-based analysis may **personalize** training plans, improve sleep habits, and help with stress management.

DOI: http://dx.doi.org/10.5772/intechopen.1014600

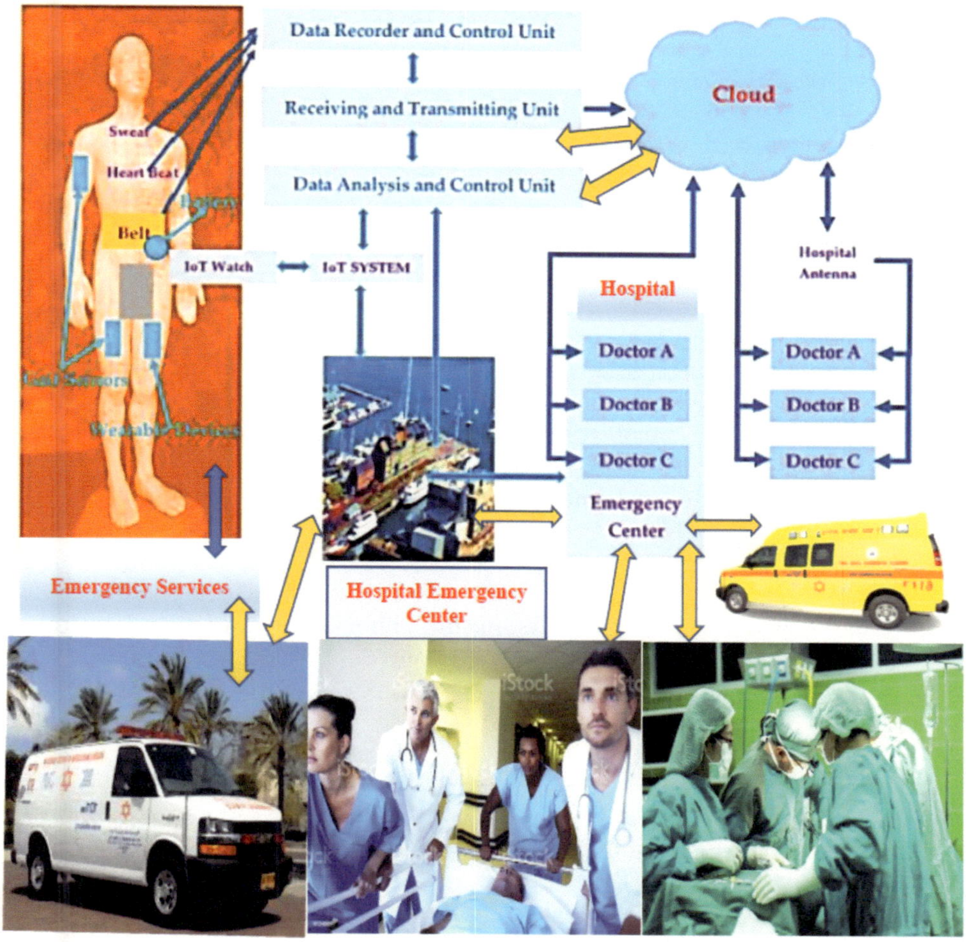

Figure 15.
Healthcare monitoring network with wearable sensors and antennas, WBAN [53] networks.

6.2 Disadvantages of wearable technologies

6.2.1 Accuracy of wearable devices

- Data may not be **accurate** compared to clinical medical devices.
- Heart rate, calories, and sleep stages can be unreliable.

6.2.2 Privacy and security issues

- Sensitive health data may be stored in the cloud.
- Data may leak, or unauthorized sharing of data may occur.

6.2.3 Short battery life of wearable devices

- Many wearables require frequent charging, limiting continuous monitoring.

- Energy harvesting units may solve these issues.

6.2.4 Cost of wearable devices

- Smart devices and watches, as well as advanced medical wearables, may be expensive.

6.2.5 Dependence on wearable devices

- Users may become overly reliant, anxious, or misinterpret data.

6.2.6 Technical limitations of wearable devices

- Limited screen size, small storage, and reliance on smartphone connectivity.
- Sensors can degrade over time.
- Some devices can be uncomfortable, cause skin irritation, or wear out quickly.
- Wearable ECGs, glucose monitors, insulin pumps, and other digital health devices improve chronic disease management.

6.3 Examples of wearable antennas

6.3.1 Textile wearable antennas

Antennas made from conductive fabrics, embroidered threads, or printed on textiles.

6.3.2 Examples

- **Patch antennas sewn into shirts**
  Used for monitoring heart rate, respiration, or transmitting biomedical data.
- **Embroidered antennas in smart shirts**
  Made from conductive yarn; used in sports performance tracking.
- **Printed flexible antennas on jackets**
  Often used for GPS or communication in outdoor and rescue operations.

6.3.3 On-body antennas for medical monitoring

Small antennas are placed on or close to the skin.

- **Wearable ECG/EEG patch antennas**
  Used in wireless medical patches (Bluetooth or NFC communication).
- **Implantable or skin-mounted UHF antennas**
  For glucose sensors, temperature patches, or health telemetry.

- **RFID adhesive antennas**
  Used for patient identification or tracking in hospitals.

6.3.4 Antennas in smartwatches and wearable devices

These antennas are embedded in consumer wearables.

- **Bluetooth and Wi–Fi antennas in smartwatches** allow wireless communication with smartphones.

- **NFC antennas in watches and rings,** used for contactless payments (Apple Watch, Oura Ring payments).

- **GPS antennas in sport watches** are used for running tracking, positioning, and outdoor navigation.

6.3.5 Wearable antennas for military and emergency use

Designed for rugged environments.

- **Body-worn communication antennas** (vest-mounted radios) for soldiers and first responders.

- **Helmet-mounted antennas (UHF/VHF)** are used in tactical communication systems.

- **Textile antennas integrated into bulletproof vests** provide long-range wireless communication without handheld antennas.

6.3.6 Antennas in smart glasses and AR/VR headsets

Used for connectivity and tracking.

- **Wi–Fi/Bluetooth antennas in smart glasses** (Ray-Ban Meta, Google Glass)

- **5G/mm Wave antennas in AR headsets** for high-speed data transfer and low-latency communication.

6.3.7 Wearable IoT antennas

Very small antennas are made for low-power devices.

- **LoRaWAN wearable antennas** used in animal tracking collars, worker safety beacons.

- **UHF RFID wearable wristbands** for tracking, access control, or health monitoring.

- **BLE beacons in badges or lanyards,** used in workplaces, hospitals, and sports events.

6.4 Safety and SAR consideration of wearable antennas

Safety and specific absorption rate (SAR) considerations of wearable antennas should be taken into account in the development of wearable antennas. The sensors and antennas presented in this chapter operate as receiving antennas, and in this case, the received power is much lower than the FCC standard, which is 1.6 W/kg averaged over 1 g of tissue. The European standard is 2 W/kg averaged over 1 g of tissue. The received power is much lower than the international standards. The sensors and antennas presented in this chapter can also operate as transmitting antennas, and in this case, the radiated power is lower than the international standards. The transmitted power is less than 20 dBm lower than cellular phones that radiate between 27 dBm and up to 30 dBm. SAR may be computed by using Eq 3. SAR measures the rate at which electromagnetic energy is absorbed by the human body when exposed to wireless devices such as mobile phones, sensors, Wi–Fi routers, and wearables.

$$\mathrm{SAR} = {}^{\sigma|E|^2}\!\big/_{\rho} \quad (3)$$

Where:
σ= tissue conductivity (S/m)
E= electric field strength (V/m)
ρ= tissue density (kg/m^3)
The wearable sensors presented in this chapter are safe and meet international standards.

6.5 Implantable antennas for medical applications

Implantable Antennas for Medical Applications are presented in the published literature; see [56, 57, 59, 60].

Implantable antennas are miniature antennas integrated into medical devices that wirelessly transmit or receive power and data from within the human body. They are used in devices like:

- Pacemakers
- Neurostimulators
- Biosensors
- Capsule endoscopes
- Wireless body area networks (WBAN)

6.5.1 Types of implantable antennas

Implantable Antennas are presented in [56–60]

- Patch antennas
- Metamatirial antennas

- Meandered slots
- Antennas on High-permittivity substrates
- Multilayer antenna structures

Materials and encapsulation used in implantable antennas must be safe for long-term implantation and prevent direct tissue contact with metal. Implantable antennas are crucial in healthcare applications that transmit or receive power and data from within the human body. Implantable antenna design for healthcare systems is a highly interdisciplinary field that employs:

- **Miniaturization**
- **Biocompatibility**
- **Low SAR and thermal safety**
- **Reliable communications via lossy tissue**

7. Conclusion

Modern wireless RF systems demand higher efficiency, wideband, and data rates. These features can be achieved by using advanced antenna technologies and wideband RF technologies. Advanced antenna technologies such as metamaterials, fractal, active antennas, MIMO antennas, notch, energy harvesting, green antennas, and several types of printed antennas were presented in this introductory chapter.

Metamaterial technology is employed to evaluate compact, efficient, wideband, and cost-effective sensors and antennas for RF devices, medical monitoring networks, healthcare sensors, and other smart gadgets. For S11 lower than -6 dB, the antenna bandwidth is around 9%. The antenna beamwidth is approximately 74°, and the antenna directivity is eight. The patch gain is around 7.7 dBi. The metamaterial patch efficiency is approximately 96%.

For S11 lower than -6 dB, the measured bandwidth of the antenna with CSSR and metallic strips is approximately 48%. The measured directivity and gain of the metamaterial antenna are 6 ± 1 dBi.

The double-layered fractal antenna bandwidth for S11 lower than -6dB is around 22%. The double-layered fractal antenna gain is 7.8 dBi.

Advanced antennas may be used in wearable sensors and devices. Wearable sensors can record electrocardiograms, measure medical parameters, analyze sweat, and monitor the health of elderly individuals and children. The wearable smart gadgets industry has experienced significant growth over the last 10 years and is employed to assist individuals in many daily life tasks.

By employing green materials in the design of communication systems and connecting an energy harvesting unit to the antenna, a green communication system may be developed. Energy harvesting systems generate green energy and can eliminate the usage of power cords and decrease the usage of batteries.

Harvesting modules can be connected to the sensors and antennas to charge the system batteries. Electronic waste, electric cables, and batteries pollute the environment. Electromagnetic energy harvesting may be a useful source of green energy.

Implanted antennas, basic technologies, and advantages are presented in this chapter.

Author details

Albert Sabban
Braude Engineering College, Karmiel, Israel

*Address all correspondence to: avi_111@netvision.net.il

References

[1] Constantine AB. Antenna Theory Analysis and Design. Fourth ed. New Jersy, USA: Wiley Sons; 2016

[2] Sabban A. Novel Wearable Antennas for Communication and Medical Systems. Boca Raton, FL, USA: Taylor & Francis Group; 2017

[3] Sabban A. Wideband RF Technologies and Antennas in Microwave Frequencies. Hoboken, NJ, USA: Wiley; 2016

[4] Sabban A. Low Visibility Antennas for Communication Systems. Boca Raton, FL, USA: Taylor & Francis; 2015

[5] Waher P. Learning internet of things, packt publishing. ISBN. 2015;**13**:97–81783553532

[6] Rajab K, Mittra R, Lanagan M Size reduction of microstrip antennas using metamaterials. In Proceedings of the 2005 IEEE APS Symposium, Washington, DC, USA, 2005

[7] Pendry JB, Holden AJ, Stewart WJ, Youngs I. Extremely low frequency plasmons in metallic mesostructures. Physical Review Letters. 1996;**76**:4773–4776. DOI: 10.1103/physrevlett.76.4773

[8] Pendry JB, Holden A, Robbins D, Stewart W. Magnetism from conductors and enhanced nonlinear phenomena. IEEE Transactions on Microwave Theory and Techniques. 1999;**47**: 2075–2084. DOI: 10.1109/22.798002

[9] Marque´s R, Mesa F, Martel J, Medina F. Comparative analysis of edge and broadside coupled split ring resonators for metamaterial design—Theory and Experiment. IEEE Transactions on Antennas and Propagation. 2003;**51**:2572–2581

[10] Marque´s R, Baena JD, Martel J, Medina F, Falcone F, Sorolla M, Martin F Novel small resonant electromagnetic particles for metamaterial and filter design. In Proceedings of the ICEAA'03, Torino, Italy, 8–12 September 2003; p. 439–442

[11] Marqués R, Martel J, Mesa F, Medina F. Left-handed-media simulation and transmission of EM waves in subwavelength split-ring-resonator-loaded metallic waveguides. Physical Review Letters. 2002;**89**:183901. DOI: 10.1103/physrevlett.89.183901

[12] Zhu J, Eleftheriades G A compact transmission-line metamaterial antenna with extended bandwidth. IEEE Antennas Wirel. Propag. Lett. 2008;**8**: 295–298. DOI: 10.1109/LAWP.2008.2010722

[13] Baena JD, Marque´s R, Martel J, Medina F Experimental results on metamaterial simulation using SRR-loaded Waveguides in Proceedings of the IEEE-AP/S International Symposium on Antennas and Propagation, Columbus, OH, USA, 22–27 June 2003; p. 106–109

[14] Marques R, Martel J, Mesa F, Medina F. A new 2D isotropic left-handed metamaterial design: Theory and experiment. Microwave and Optical Technology Letters. 2002;**35**:405–408. DOI: 10.1002/mop.10620

[15] Sabban A. New compact wearable metamaterials circular patch antennas

for IoT, medical and 5G applications. Applied System Innovation. 2020;**3**:42. DOI: 10.3390/asi3040042

[16] Sabban A. Small wearable metamaterials antennas for medical systems. Applied Computational Electromagnetics Society Journal. 2016;**31**:434–443

[17] Fujimoto K. Antenna for Small Mobile Terminals. Norwood MA, USA: Artech House; 2018

[18] Sabban A. Wearable antenna measurements in vicinity of human body. Wireless Engineering and Technology. 2016;7:97–104. DOI: 10.4236/wet.2016.73010

[19] Sabban A. Active compact wearable body area networks for wireless communication, medical and IOT applications. Applied System Innovation. 2018;**1**:46. DOI: 10.3390/asi1040046

[20] Mandelbrot BB. "The Fractal Geometry of Nature". N.Y USA: W.H. Freeman and Company; 1983

[21] Rizzoli V, Costanzo A, Spadoni P Computer-aided design of ultra-wideband active antennas by means of a new figure of merit. IEEE Microw. Wirel. Compon. Lett. 2008;**18**: 290–292. DOI: 10.1109/LMWC. 2008.918971

[22] James JR, Hall PS, Wood C. Microstrip Antenna Theory and Design. London, UK: Institution of Engineering and Technology (IET); 1981

[23] Sabban A Dual polarized dipole wearable antenna. U.S. Patent 8,203,497, 2012

[24] Falkoner FJ. The Geometry of Fractal Sets. Cambridge UK: Cambridge Univ. Press; 1990

[25] Minin I. "Microwave and Millimeter Wave Technologies from Photonic Bandgap Devices to Antenna and Applications", INTECH, Chapter 16, Fractal Antenna Application by Mircea V. Intech, London UK: Rusu and Roman Baican; 2010 March. P. 351–382

[26] Rusu MV, Hirvonen M, Rahimi H, Enoksson P, Rusu C, Pesonen N, Vermesan O, Rustad H, "Minkovsky fractal microstrip antenna for RFID tags", Proc. EuMW2008 Symposium, Amsterdam, October 2008

[27] Rahimi H, Rusu M, Enoksson P, Sandström D, Rusu C, Small patch antenna based on fractal design for wireless sensors, MME07, 18th Workshop on Micromachining, Micromechanics, and Microsystems 16-18 sept. 2007, Portugal

[28] Mandelbrot BB. How long is the coast of Britain? statistical self-similarity and mandelbrot, B how long is the coast of britain? statistical self-similarity and fractional dimension. Science (New York, N.Y.). 1967;**156**:636–638

[29] Chirwa L, Hammond P, Roy S, Cumming DRS. Electromagnetic radiation from ingested sources in the human intestine between 150 MHz and 1.2 GHz. IEEE Transactions on Biomedical Engineering. 2003;**50**:484–492. DOI: 10.1109/tbme.2003.809474

[30] Werber D, Schwentner A, Biebl EM. Investigation of RF transmission properties of human

tissues. Advances in Radio Science. 2006;**4**:357–360. DOI: 10.5194/ars-4-357-2006

[31] Gupta B, Sankaralingam S, Dhar S Development of wearable and implantable antennas in the last decade: A review. In Proceedings of the 2010 10th Mediterranean Microwave Symposium, Guzelyurt, Cyprus, 25–27 August 2010; p. 251–267

[32] Thalmann T, Popovic Z, Notaros BM, Mosig JR Investigation and design of a multi-band wearable antenna. In Proceedings of the 3rd European Conference on Antennas and Propagation, EuCAP 2009, Berlin, Germany, 23–27 March 2009; p. 462–465

[33] Salonen P, Rahmat-Samii Y, Kivikoski M Wearable antennas in the vicinity of human body. In Proceedings of the IEEE Antennas and Propagation Society Symposium, Monterey, CA, USA, 20–25 June 2004; p. 467–470

[34] Kellomäki T, Heikkinen J, Kivikoski M Wearable antennas for FM reception. In Proceedings of the 2006 First European Conference on Antennas and Propagation, Nice, France, 6–10 November 2006; p. 1–6

[35] Sabban A Microstrip antenna and antenna arrays. U.S. Patent 4,623,893, 1986

[36] Wheeler HA. Small antennas. IEEE Transactions on Antennas and Propagation. 1975;**23**:462–469

[37] Jamil F, Ahmad S, Iqbal N, Kim D. Towards a remote monitoring of patient vital signs based on IoT-based blockchain integrity management platforms in smart hospitals. Sensors. 2020;**20**:2195. DOI: 10.3390/s20082195

[38] Lin J, Itoh T. Active integrated antennas. IEEE Transactions on Microwave Theory and Techniques. 1994;**42**:2186–2194. DOI: 10.1109/22.339741

[39] Mortazwi A, Itoh T, Harvey J. Active Antennas and Quasi-Optical Arrays. New York, NY, USA: John Wiley & Sons; 1998

[40] Jacobsen S, Klemetsen O. Improved detectability in medical microwave radio-thermometers as obtained by active antennas. IEEE Transactions on Biomedical Engineering. 2008;**55**:2778–2785. DOI: 10.1109/tbme.2008.2002156

[41] Jacobsen S, Klemetsen Ø. Active antennas in medical microwave radiometry. Electronics Letters. 2007;**43**:606. DOI: 10.1049/el:20070577

[42] Ellingson SW, Simonetti JH, Patterson CD. Design and evaluation of an active antenna for a 29–47 MHz radio telescope array. IEEE Transactions on Antennas and Propagation. 2007;**55**:826–831

[43] Segovia-Vargas D, Castro-Galan D, Munoz LEG, González-Posadas V. Broadband active receiving patch with resistive equalization. IEEE Transactions on Microwave Theory and Techniques. 2008;**56**:56–64. DOI: 10.1109/TMTT.2007.912008

[44] Paradiso JA, Starner T Energy scavenging for mobile and wireless electronics. IEEE Pervasive Comput. 2005;**4**:18–27

[45] Valenta CR, Durgin GD. Harvesting wireless power: Survey of energy-harvester conversion efficiency in far-field, wireless power transfer

systems. IEEE Microwave Magazine. 2014;**15**:108–120

[46] Nintanavongsa P, Muncuk U, Lewis DR, Chowdhury KR Design optimization and implementation for RF energy harvesting circuits. IEEE J. Emerg. Sel. Top. Circuits Syst. 2012;**2**:24–33

[47] Devi KKA, Sadasivam S, Din NM, Chakrabarthy CK Design of a 377 Ω patch antenna for ambient RF energy harvesting at downlink frequency of GSM 900. In Proceedings of the 17th Asia Pacific Conference on Communications (APCC '11), Sabah, Malaysia, 2–5 October 2011; p. 492–495

[48] Sabban A. "Wearable Communication Systems and Antennas, Second Edition: Design, Efficiency, and Miniaturization Techniques". Bristol UK: IOP Publication; 2022 June

[49] Goudarzi A, et al. A Survey on IoT-Enabled Smart Grids: Emerging, Applications, Challenges, and Outlook, MDPI Energies. Journal paper, MDPIEnergies Journal. September 2022. DOI: 10.3390/en15196984

[50] Waseem M, et al. Incorporation of blockchain technology for different smar grid applications: Architecture, prospects, and challenges. MDPI Energies January. 2023. DOI: 10.3390/en16020820

[51] Sabban A. New wideband printed antennas for medical applications. IEEE Transactions on Antennas and Propagation. 2013;**61**:84–91. DOI: 10.1109/tap.2012.2214993

[52] Sabban A.Novel meta-fractal wearable sensors and antennas for medical, communication, 5G, and IoT applications. Fractal MDPI Journal. 2024;**8**(2):100

[53] Sabban A. Wearable circular polarized antennas for health care, 5G, energy harvesting, and IoT systems. Electronics 2022. 2022;**11**(3):427. DOI: 10.3390/electronics11030427

[54] Lamkaddem A, Yousfi AE, Vargas DS. Perspective chapter: Challenges and future trends in wireless power transfer for biomedical implants. The Challenges of Energy Harvesting. IntechOpen. 2025. DOI: 10.5772/intechopen.1009236

[55] Elabdi A, Ayachi ME, Rahmoun M, Lamkaddem A, "A review of reconfigurable antenna types and mechanisms for wireless communications: IoT applications," International Conference on Circuit, Systems and Communication (ICCSC), Fez, Morocco, 2025, pp. 1–8, DOI: 10.1109/ICCSC66714.2025.11135324

[56] Lamkaddem AEY, González-Posadas V, Segovia-Vargas D. Miniaturized dual band implantable antenna for implanted biomedical devices. IEEE Access. 2024;**12**: 15026–15036. DOI: 10.1109/ACCESS.2024.3357488

[57] Ahmad S, Yousfi AE, Lamkaddem A, Vargas DS, "A low-profile CPW-fed circularly polarized antenna for biomedical applications," 2022 IEEE International Symposium on Antennas and Propagation and USNC-URSI Radio Science Meeting (AP-S/URSI), Denver, CO, USA, 2022, pp. 107–108, DOI: 10.1109/AP-S/USNC-URSI47032.2022.9886402

[58] ADS Momentum Software, Keysightt. Available from: https://www.keysight.com/en/pc-1297113/advanced-design-system-ads?cc=IL&lc=eng [Accessed: 2018-January-03]

[59] Soliman MM, et al. Review on medical implantable antenna technology and imminent research challenges. Sensors. 2021;**21**(9):3163. DOI: 10.3390/s21093163

[60] Ahmed Z, Zaki A, et al. Design and modelling of ultra-compact wideband implantable antenna for wireless ISM band. Bioengineering. 2023;**10**(2):216. DOI: 10.3390/bioengineering10020216

Chapter 2

Review and Design of Antenna Technologies for Wearable Devices: Dual-Mode Antenna Design for Wireless Body Area Networks

Dairoku Muramatsu

Abstract

This chapter reviews antenna technologies for wearable devices, with an emphasis on IEEE 802.15.6 wireless body area networks (WBANs). We clarify a dual-mode framework in which electro-quasistatic human body communication (HBC, tens of megahertz) complements 2.4 GHz far-field links to deliver robust, energy-aware on-/off-body connectivity. The presentation is tutorial in style: we outline planar-antenna fundamentals and a practical design flow for a microstrip patch at 2.4 GHz, explain the physical mechanisms that govern HBC, and use analysis-based illustrations (including a full-body field-distribution example) to build shared intuition. On this foundation, we organize concise design guidelines for dual-mode platforms – covering geometry and spacing, feed routing and common-mode control, ground shaping and shielding, electrode/contact design for HBC, coexistence across radios, power and safety planning, and a concept-level verification matrix tied to mode-selection policy. Read together, these elements form a repeatable workflow for allocating roles between on- and off-body links, anticipating body-proximity effects, and documenting acceptance criteria that shorten iteration cycles. We close with an outlook on system integration and validation to translate the framework into dependable WBAN products.

Keywords: wearable antenna, wireless body area network, physical layer, dual-mode, electromagnetic field analysis

1. Introduction

Due to rapid advances in electronics, information devices have evolved to become so small, lightweight, and high-performance that users scarcely notice wearing them. "Wearables," once showcased at events such as the International Consumer Electronics Show in the United States and the Combined Exhibition of Advanced

Technologies in Japan, are no longer futuristic; they have become familiar consumer items readily available at electronics retailers. A key feature of wearable devices is their ability to deliver a wide variety of functions, spanning healthcare, fitness, entertainment, and industrial applications. One of the foundational technologies enabling these functions is the communication system that interconnects wearable devices with each other and with external devices. In this chapter, we provide a concept-first overview of networking and antenna technologies for wearable devices and supply concrete, analysis-based tutorials without relying on experimental datasets.

2. Wearable healthcare and wireless body area network technology

Information and communication devices have evolved from mobile to wearable, and it has become common to implement biosignal-sensing functions – such as heart rate and acceleration – in devices exemplified by smartwatches. Great expectations are being placed on "integrated healthcare in a wearable environment" (see **Figure 1**) [1], where wearable devices continuously measure and collect a user's state, and a medical server monitors the data to maintain health and prevent severe illness. Numerous wearable devices (sensors) attached to the user are interconnected through on-body communication on the human body and eventually connect via off-body communication to devices located away from the user. In this way, the biosignals acquired are transmitted continuously to a medical server to monitor health conditions.

As the short-range wireless technology surrounding the user that supports such a healthcare system, the wireless body area network (WBAN) has been standardized in IEEE 802.15.6. The standard primarily targets medical healthcare applications and, as summarized in **Table 1**, features three different physical layers: human body communication (HBC) in the 21 MHz band; narrow band wireless communication (NBWC) using, for example, the 400 MHz, 920 MHz, and 2.4 GHz bands; and ultra-wideband (UWB) using 3.4–4.8 GHz and 6.25–10.25 GHz – while employing a common medium access control layer [2].

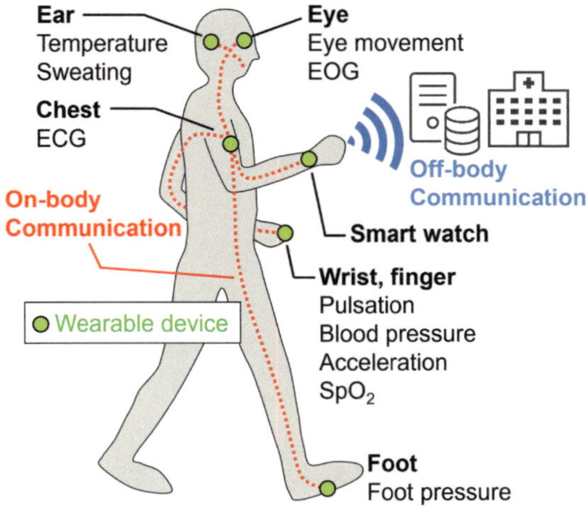

Figure 1.
Integrated healthcare in a wearable environment.

Usage	**On-body**	**Off-body**		
PHY	HBC	NBWC		UWB
Frequency band	21 MHz band	402–405 MHz	Implantable device (Universal)	3.4–4.8GHz 6.25–10.25 GHz, regulated by each country
		420–450 MHz	Medical telemetry (Japan)	
		863–870 MHz	ISM band (EU)	
		902–928 MHz	ISM band (USA)	
		950–956 MHz	Specified low power radio	
		2.36 – 2.4 GHz	Medical device(USA)	
		2.4 GHz band	ISM band (Universal)	
Modulation		D-PSK, GFSK		OOK, D-PSK
Data rate	<1.3 Mbps	<970 kbps		>12 Mbps
MAC	All common			

Table 1.
Technical specifications of IEEE 802.15.6 WBAN.

In this section, we focus on the 2.4 GHz-band narrow band wireless communication and the 21 MHz-band HBC within WBAN technology, and we describe the operating principles and a design example of a patch antenna, the transmission mechanism of HBC, and a development example of a dual-mode wearable antenna that covers multiple physical layers.

2.1 Scope and methods of this review

We synthesize standards documents and representative peer-reviewed studies to clarify the underlying physical mechanisms and their implications for wearable antenna design in WBANs. The presentation is tutorial in style: concise analytical expressions and illustrative electromagnetic analyses are used where they sharpen intuition. Figures emphasize conceptual understanding and design flow, and the framework developed here structures the subsequent sections and the concluding guidelines.

2.2 Dual-mode concepts for WBANs

In this chapter, dual-mode denotes the coordinated use of two distinct coupling mechanisms on or around the body: (1) electro-quasistatic near-field coupling at tens of megahertz for on-body links (HBC), where fields are confined around tissue and decay rapidly with distance; and (2) far-field radiation in the 2.4 GHz ISM band for off-body links, where proximity to the body perturbs matching, efficiency, and patterns. We outline three archetypes: (1) a single-feed structure with a shared ground, (2) switchable paths using a common radiator, and (3) dual-port architectures – and discuss qualitative trade-offs in volume, isolation, and robustness. This conceptual framework guides the assignment of roles to each mode in system design.

3. Wearable devices and planar antenna technology

Antennas are among the most important components in wireless communication systems. By coupling information to and from space in the form of electromagnetic waves, antennas enable wireless communication and are used in a wide variety of applications – smartphones, TV, radio, amateur radio – with diverse shapes, sizes, and implementation methods. The first structures that come to mind are fundamental antennas such as dipoles and monopoles. These are easy to design to meet required specifications (e.g. resonant frequency) and can provide relatively high antenna gain. However, because they are physically large and three-dimensional, integration into compact wearable devices is generally difficult. Consequently, planar antennas are prime candidates for implementation within the limited space of wearable devices [3].

A representative planar antenna is the patch antenna (microstrip antenna in the strict sense) shown in **Figure 2** [4]. A patch antenna is a planar structure with a radiating element (patch conductor) and a ground plane (ground conductor) placed on opposite sides of a dielectric substrate, fed between the radiating element and the ground. Depending on the shape of the radiating element, there are rectangular and circular patch antennas, each with distinctive polarization characteristics, among others. The structure is equivalent to a microstrip line open-circuited at both ends; it is a type of open resonator that resonates when the patch length L is an integer multiple of the half-wavelength [5]. In particular, the patch antenna employs a dielectric substrate with small relative permittivity ε_r, and sets the patch width W and the substrate thickness h to values slightly larger than those used purely as a resonator (while still electrically thin), thereby increasing radiation and using the microstrip line (i.e. the patch) as the radiating element.

The characteristics of patch antennas include: (1) relatively high antenna gain; (2) the possibility of microstrip-line feeding, enabling integration of the radiating and feeding structures; and (3) low-cost, high-volume production

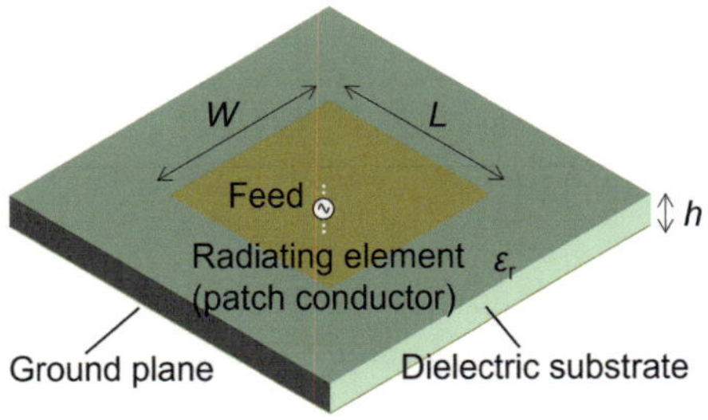

Figure 2.
Basic structure of a rectangular patch antenna.

using printed-circuit technology. In the next subsection, we present a concrete design and analysis example of a microstrip-fed rectangular patch antenna.

4. Design and analysis example of a rectangular patch antenna

As a representative planar antenna with a simple structure, we present a design and analysis example of a microstrip-fed rectangular patch antenna [6]. The material constants and dimensions are shown in **Figure 3**: the substrate relative permittivity ε_r, substrate thickness h, substrate length L_d, substrate width Wd, patch length L_p, patch width W_p, microstrip feedline width W_m, gap length L_g, and gap width W_g. These dimensions can be approximately calculated from ε_r, h, and the target resonant frequency f_0 [6]. For example, assuming a dielectric substrate with $\varepsilon_r = 3.5$ and $h = 5$ mm, and designing the antenna to resonate at $f_0 = 2.4$ GHz, the dimensions in **Figure 3** become $L_d = 66.9$ mm, $W_d = 77.4$ mm, $L_p = 31.1$ mm, $W_p = 41.6$ mm, $W_m = 10.4$ mm, $L_g = 10.1$ mm, and $W_g = 0.94$ mm. Numerous simple web-based calculators for patch-antenna baseline dimensions yield nearly identical results. Note, however, that the feed position parameter L_g was determined through the numerical electromagnetic analysis described below to achieve input-impedance matching.

To confirm the designed antenna's resonance, we calculated antenna characteristics using XFdtd (Remcom Inc.), a numerical EM simulator based on the finite-difference time-domain (FDTD) method. **Figure 4** shows the frequency response of the reflection coefficient S_{11} of the patch antenna. At a resonant frequency of 2.42 GHz, $S_{11} = -17.5$ dB was obtained, demonstrating good resonance. In addition, the radiation efficiency at 2.4 GHz was 94%, indicating excellent radiation performance for the designed patch antenna.

Thus, patch antennas can be designed more easily than more complex antenna structures. As noted above, this is because patch antennas are based on the principle of an open resonator, and approximate formulae agree relatively well with numerical analysis. On the contrary, to achieve superior resonance characteristics at the desired center frequency and bandwidth, small dimensional variations and dielectric losses may have significant effects; hence, it is recommended to combine approximate design methods with electromagnetic analysis [7–9].

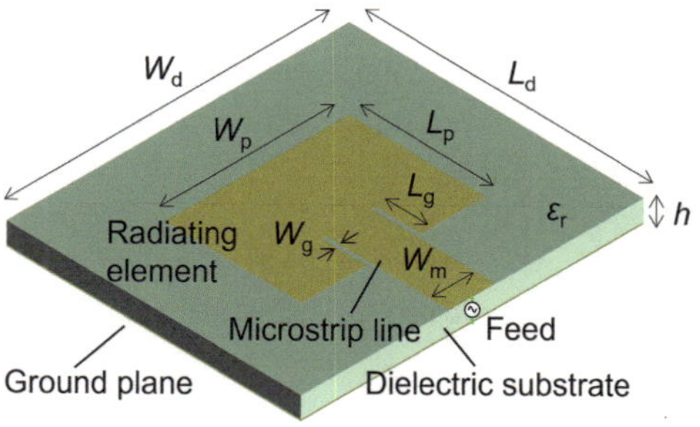

Figure 3.
Basic structure of a microstrip-fed rectangular patch antenna.

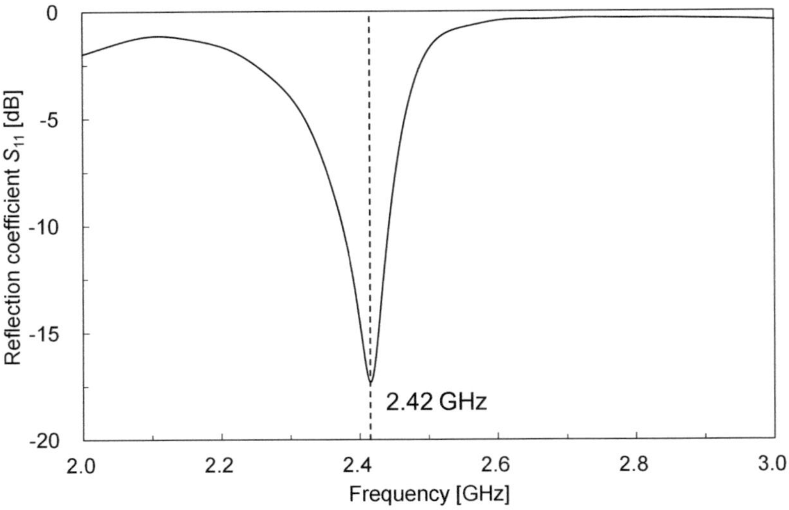

Figure 4.
Resonance characteristics of the rectangular patch antenna. Resonance characteristics of the rectangular patch antenna.

5. HBC technology

Wireless systems intended for use around the user commonly employ, as noted earlier, Bluetooth Low Energy (BLE) in the 2.4 GHz band, low-power radio at 315 MHz, and specified low-power radio at 400 MHz. It is generally known that the higher the frequency, the more electromagnetic energy is absorbed by the human body; in these bands (315 MHz to 2.4 GHz), the presence of the body acts as an obstacle [10]. In this way, the human body is a nuisance in wireless communication because it attenuates electromagnetic waves and degrades communication. There is, however, the concept of treating the body itself as a transmission path for high-frequency signals and exploiting it [11]. This is the technology called HBC, a type of wireless communication in which high-frequency signals are coupled to and from the body via electrodes (antennas) in contact with or in proximity to the body.

In HBC, currents flowing on the body surface and the near-field electric field formed around the body contribute to communication. Electrically, the human body exhibits both conductive and dielectric properties; when a high-frequency signal is excited into the body via electrodes, a high-frequency current path is formed on the body surface. Moreover, at frequencies of several tens of megahertz and below, a near-field (quasi-electrostatic) field arises at the interface between the body and the surrounding space (air), dominating over the radiated far field [12–14]. Because the near field decays exponentially with distance, electromagnetic fields scarcely leak into the surrounding space during signal transmission. This enables communication with high confidentiality and minimal electromagnetic noise. In addition, because the communication distance is limited, it is potentially possible to operate at lower power than conventional wireless technologies that use airborne electromagnetic waves. Moreover, since the transmission path is established by "touching" the target person or object, it can be used to realize intuitive human interfaces that leverage user actions.

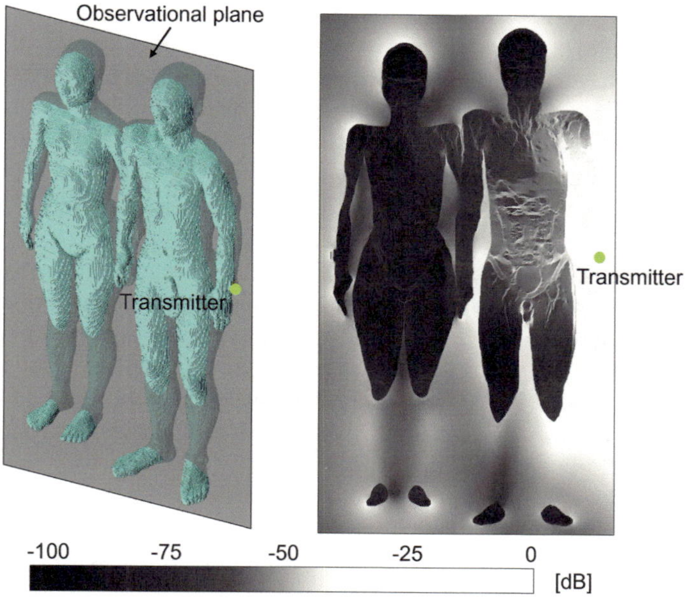

Figure 5.
Example calculation of the electric-field distribution around the human body in human body communication.

Figure 5 shows an example of the electric-field distribution around the human body in HBC, computed via EM analysis using two full-body human models [15]. In this example, a 10 MHz signal is excited from an electrode worn on the right-hand-side male model's wrist [16]. The field is distributed only around the human body, and strong fields are also observed around the hand-held female model, who is connected by hand. Conversely, the field strength rapidly decreases at positions farther away from the body.

Although nearly 30 years have passed since HBC research began, its recognition remains limited, and it has not yet seen large-scale practical deployment. One reason is that it is still an evolving technology lacking a well-systematized design methodology. In particular, the design of communication electrodes that interface between the body and equipment is critical; for example, the improvement of transmission characteristics using impedance-matching electrodes has been investigated [17]. In the next subsection, we present a more advanced electrode – a dual-mode antenna that integrates a patch antenna with an HBC electrode.

6. Design guidelines for wearable dual-mode platforms

This section consolidates design implications from the concepts and tutorials in Sections 2–5 (HBC at tens of megahertz and 2.4 GHz far-field radiation) into concise, actionable guidance for device engineers.

Guideline 1 – Roles and mode selection.

Plan HBC for on-body hops/relays and 2.4 GHz for off-body access and bursts; define simple RSSI/SNR-based thresholds with hysteresis so the device switches modes predictably rather than flapping.

Guideline 2 – Geometry, spacing, and packaging.

Maintain a thin, low- ε spacer between the radiator and the body where feasible; co-design the mechanical stack-up (textile/foam, curvature, mounting) with RF to ensure resonance and efficiency remain stable during wear.

Guideline 3 – Feed routing and common-mode control.

Short, well-referenced feed paths, compact chokes/baluns, and ground-adjacent cable routing limit unintended common-mode currents into the body and housing.

Guideline 4 – Ground shaping and local shielding.

Use slots, notches, or partial grounds to reduce body-facing current density while keeping the radiator effective. When industrial design necessitates proximity to the skin, consider using local shields or liners and verify their impact analytically.

Guideline 5 – HBC electrode and contact design.

Stabilize contact impedance with compliant surfaces or textures; ensure clear return paths and chassis referencing; consider guards around electrodes to reduce stray coupling to off-body objects.

Guideline 6 – Coexistence across radios.

Plan BLE/UWB/NFC/etc. coexistence early provide basic low-pass behavior for HBC paths to reject GHz energy; coordinate band-limiting and TX/RX blanking windows during mode switching.

Guideline 7 – Power budgeting and safety.

Exploit HBC for low-energy control/telemetry, reserving 2.4 GHz for data bursts; document duty-cycle envelopes and power-limit tables versus placement; maintain SAR margins at GHz and ensure safe electrode operation at MHz.

Guideline 8 – Verification matrix and documentation.

Define a concept-level test matrix (free-space vs. on-body; wrist/arm/torso; postures; curvature; cable orientations; housing materials; EMC/coexistence checks) and establish acceptance criteria aligned with the mode-selection policy; document this in the product documentation to ensure reproducibility.

7. Conclusions

This chapter articulates a coherent dual-mode framework in which electro-quasistatic HBC at tens of megahertz complements 2.4 GHz far-field links within IEEE 802.15.6 WBANs. By separating the roles of on-body and off-body connectivity and by clarifying the underlying physical mechanisms, the chapter provides a common vocabulary for allocating functions across the two modes and for reasoning about their interactions with the human body and the device enclosure.

On this foundation, we present a tutorial path for planar-antenna design at 2.4 GHz, along with an analysis-based explanation of HBC field behavior around and between bodies. The combination is intentional: the patch-antenna tutorial captures the levers available to the GHz-band designer under proximity constraints, while the HBC exposition explicitly explains why near-field coupling behaves differently from conventional radiation and how that behavior can be harnessed for robust on-body links. The result is a concept-first view that engineers can apply across device form factors, mounting locations, and industrial-design constraints.

We then translated the concepts into design guidance that emphasizes end-to-end co-design: mechanical stack-up and spacing to temper detuning; feed routing and common-mode control to limit unintended currents; ground shaping and local shielding to manage body-facing fields; electrode and contact planning for stable HBC coupling; coexistence planning across radios; power budgeting and safety envelopes; and a verification matrix that ties physical testing to mode-selection policy. Read together, these elements describe a practical workflow – define roles, stabilize the electromagnetic platform, plan coexistence, and document acceptance criteria – that can shorten iteration cycles and improve predictability in real products.

Viewed at the system level, a role-aware dual-mode platform mitigates coverage nulls, reduces the average energy per delivered bit for routine on-body traffic, and supports privacy by keeping selected exchanges confined to near-field coupling on the body while reserving the GHz band for interoperability and bursts. The framework also clarifies trade-offs: volume versus isolation in the RF architecture, spacer thickness versus comfort and esthetics in mechanical design, and safety margins versus performance under diverse wear conditions. Making these trade-offs explicit at design time is crucial for reproducible outcomes across users, placements, and postures.

Finally, we delineated the scope and outlook. This chapter is intentionally analysis-driven and tutorial in style; comprehensive prototype fabrication and empirical evaluation – encompassing input characteristics, patterns, and efficiency under proximity, on-body transfer behavior across placements and postures, bending and aging effects, and exposure metrics – are reserved for separate research outputs. Future work will also benefit from unified models that co-predict gigahertz detuning and megahertz coupling, and from firmware policies that adapt mode selection to per-user wear conditions. Together, these efforts will test and refine the framework advanced here and quantify its benefits in field use.

Acknowledgments

This work was partially supported by JSPS KAKENHI Grant No. 23K25250. The author wishes to thank Ms. Y. Sasaki for her assistance with the electromagnetic field simulations.

Conflict of interest

The authors declare no conflicts of interest.

Author details

Dairoku Muramatsu
University of Electro-Communications, Chofu, Japan

*Address all correspondence to: dmuramatsu@uec.ac.jp

References

[1] Vijayan V, Connolly JP, Condell J, McKelvey N, Gardiner P. Review of wearable devices and data collection considerations for connected health. Sensors. 2021;**21**(16):16. DOI: 10.3390/s21165678

[2] IEEE Computer Society. IEEE Standard for Local and Metropolitan Area Networks – Part 15.6: Wireless Body Area Networks. New York: IEEE; 2012

[3] Richards W, Lo Y, Harrison D. An improved theory for microstrip antennas and applications. IEEE Transactions on Antennas and Propagation. 1981;**29**(1):38–46. DOI: 10.1109/TAP.1981.1142529

[4] Deschamps GA Microstrip microwave antennas. In: Proceedings of the 3rd USAF Symposium on Antennas; 1953

[5] Pozar DM. Microwave Engineering. 3rd ed. Hoboken: John Wiley & Sons; 2005

[6] Carver KR, Mink JW. Microstrip antenna technology. IEEE Transactions on Antennas and Propagation. 1981;**29**(1):2–24. DOI: 10.1109/TAP.1981.1142523

[7] Aina TS. Investigation on performance of microstrip patch antenna for a practical wireless local area network (WLAN) application. International Journal for Research in Applied Science & Engineering Technology. 2022;**10**(1):221–228. DOI: 10.22214/ijraset.2022.40175

[8] Abdulhussein AM, Khidhir AH, Naser AA. 2.4 GHz microstrip patch antenna for S-band wireless communications. Journal of Physics: Conference Series. 2021;**2114**(1):012029. DOI: 10.1088/1742-6596/2114/1/012029

[9] Asokan V, Thilagam S, Kumar KV Design and analysis of microstrip patch antenna for 2.4 GHz ISM band and WLAN application. In: Proceedings of the 2015 2nd International Conference on Electronics and Communication Systems (ICECS); February 26-27, 2015; Coimbatore, India. New York: IEEE; 2015. pp. 1114–1118

[10] Murakuni E, Higaki N, Shiba K. Analysis of human body electric field distribution using the transmitting antenna in an implantable device. IEICE Technical Report. 2009;**33**(15):13–18

[11] Zimmerman TG Personal Area Networks (PAN): Near-field Intra-body Communication [thesis]. Cambridge: MIT Media Laboratory; 1995

[12] Kato Y New type sensing technology using evanescent electromagnetic field area. In: Proceedings of the International Conference on Electronics Packaging (ICEP 2012); April 18-20, 2012; Tokyo, Japan. pp. 727–730

[13] Noda A, Shinoda H. Selective wireless power transmission through high-Q flat waveguide-ring resonator on 2-D waveguide sheet. IEEE Transactions on Microwave Theory & Techniques. 2011;**59**(8):2158–2167. DOI: 10.1109/TMTT.2011.2151830

[14] Datta A, Yang D, Maity S, et al. Touchscreen communication (ToSCom): Electro-quasistatic body

communication during touch sensing. Communications Engineering. 2025;**4**:56

[15] Nagaoka T, Watanabe S, Sakurai K, Kunieda E, Watanabe S. Development of realistic high-resolution whole-body voxel models of Japanese adult male and female of average height and weight and application of models to radio-frequency electromagnetic-field dosimetry. Physics in Medicine & Biology. 2004;**49**:1–15. DOI: 10.1088/0031-9155/49/1/001

[16] Muramatsu D, Koshiji F, Koshiji K, Sasaki K Transmission characteristics of wearable devices between two human bodies. In: Proceedings of the 1st IEEE Global Conference on Consumer Electronics (GCCE); October 2-5, 2012; Tokyo, Japan. New York: IEEE; 2012. p. 172

[17] Muramatsu D, Koshiji F, Koshiji K, Sasaki K. Analytical and experimental studies on human body communication between wristwatch and handheld device using muscle homogenous phantom at 10 MHz. Sensors and Materials. 2014;**26**(8): 581–589. DOI: 10.18494/SAM.2014.1021

Chapter 3

Review of Microwave (RF) Energy Harvesting Techniques

V. Lavanya, C. Rimmya, S. Piramasubramanian and Madhan M. Ganesh

Abstract

The advancement in wireless technology has greatly expanded its applications to different fields. Thus, the usage of electronic components and the power required to charge them have also increased. To overcome the dependency on batteries, ambient energy can be collected, converted, and finally utilized, which is known as energy harvesting. The fundamental concept behind wireless energy harvesting (WEH) is the conversion of energy from sources such as radio frequency (RF) signals, electromagnetic waves, and mechanical vibrations into electrical energy that may be used to power electronic equipment. RF energy harvesting refers to converting energy from the electromagnetic field into electrical energy. The history and concept behind RF energy harvesting are discussed in this chapter. It also provides an overview of the RF energy harvesting architecture system and rectenna system. Antenna design types, metamaterial- and metasurface-based antennas, and applications of rectenna systems in Internet of Things (IoT), medical, and communication are reviewed. A glimpse into Simultaneous Wireless Information and Power Transfer (SWIPT) is provided, and the chapter concludes with challenges.

Keywords: electromagnetic energy harvesting, rectenna, wireless power transfer, IoT, simultaneous wireless information and power transfer (SWIPT)

1. Introduction

The demand for energy in the latest electronic devices is increasing day by day. Batteries are the only primary energy source for the present power electronic devices. However, they have challenges like limited lifespan, periodic replacement, and environmental concerns. Hence, the need for alternative power generation techniques like energy harvesting, which ensures a continuous power supply, particularly for devices with low power, is being developed. The technique is to harvest energy from ambient sources and convert it into usable direct current (DC).

The energy harvesting or energy scavenging involves collecting, converting, and utilizing energy from naturally available environmental sources such as solar, thermal, kinetic, and radio frequency (RF) radiation. These ambient sources have been widely explored for enabling energy-autonomous operation of Internet of Things (IoT) systems and machine-to-machine (M2M) communication networks [1]. **Figure 1** compares stored battery energy and harvested ambient energy with respect to time [2]. The accompanying **Table 1** shows output voltage from various ambient energy sources like thermal, wind, solar, vibration, and electromagnetics (EMs) [3]. In EMs, the nature of the source is continuous, and an output voltage of about 3–4 V can be obtained.

Among these ambient sources, RF energy harvesting (RFEH) stands out due to its ability to scavenge wasted or unused RF energy by the end user from the environment. It converts the energy into usable DC power – both day and night, and across indoor and outdoor environments [1, 4]. RFEH systems generally rely on a rectenna, a combination of a receiving antenna and a rectifying circuit, which captures the EM energy and converts it into DC power [4, 5].

This chapter explores the historical overview and design principles of antennas for RFEH, as well as their applications in IoT, medical devices, and communication systems. Additionally, rectenna architectures, rectifier topologies, and relevant state-of-the-art advancements are analyzed to provide a comprehensive understanding of modern RFEH systems.

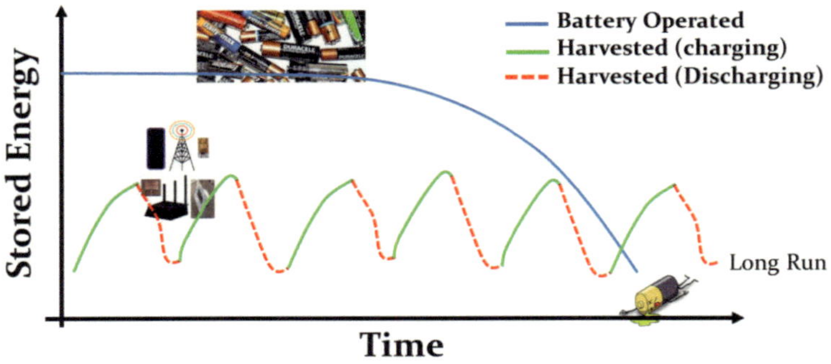

Figure 1.
Comparison of stored energy versus time among battery-operated (blue line), harvested charging (green line) and discharging (red dashed line).

Ambient sources	Power density	Availability	Output voltage
Thermal	40μ–100 μW/cm^2	Continuous	10–100 MW
Wind	3.5 mW/cm^2	Continuous	100 MW and above
Solar	1,500 μW/cm^2	Day time	0.5–1 V
Vibration	3.8–500 μW/cm^2	Activity dependent	10–25 V
Electromagnetic (EM)	0.02–1 μW/cm^2	Continuous	3–4 V

Table 1.
Output voltage from various ambient energy sources.

2. Historical overview

In 1864, James Clerk Maxwell predicted the existence of radio waves as a form of energy. After two decades, his theory was experimentally proven by Heinrich Hertz, who successfully transmitted and received radio waves. In the late nineteenth century, Guglielmo Marconi marked the beginning of practical wireless communication by publicly transmitting signals [1]. Around the same period, Nikola Tesla conducted experiments in wireless power transfer (WPT). His work addressed both near-field and far-field transmission methods. Tesla utilized electric fields for near-field WPT and envisioned a global wireless energy system. However, due to the technological limitations of the time – particularly the lack of efficient high-power generation, funding, and transmission methods – further research in WPT was unable to continue [6].

After World War II, wireless power experiments rose with the development of microwave technology and high-power vacuum tubes [6]. These innovations made it feasible to beam power over long distances. In the 1960s, William C. Brown introduced the first rectifying antenna (rectenna) and demonstrated microwave power transmission by successfully powering a flying drone in 1964 [1].

Similarly, in 1968, the Solar Power Satellite (SPS) was proposed, which involved gathering solar energy in space, transforming it into microwave energy, and transmitting it wirelessly to Earth. Upon reception, rectenna arrays would then convert the microwave energy into usable DC power [1]. In recent years, rectennas have extended their contribution to IoT, medical, and communication fields.

3. Principle behind RFEH

Wireless energy transfer (WET) and RFEH primarily depend on the principles of EM induction and EM resonance.

- **EM induction** is the process in which a time-varying magnetic field induces an electric current in a nearby coil. This principle enables energy transfer in systems where the transmitter and receiver are closely aligned.

- **EM resonance** occurs when two coils are tuned to the same natural frequency, allowing efficient energy transfer over a longer distance. It enhances coupling strength even with spatial separation, reducing sensitivity to alignment.

RFEH is classified into:

Far-field (radiative) RFEH:

It uses EM wave propagation (typically in the 300 MHz to 300 GHz range) to transfer energy over long distances with minimal infrastructure. However, the low power density and poor safety concerns due to potential radiation that occurs beyond RF safety limits make it unsuitable in human environments [7]. It is ideal for applications such as satellite systems, defense, and industrial wireless powering.

Near-field (non-radiative) RFEH:

It uses EM wave propagation to transfer energy over short distances. It offers high power density and safer operations compared to far-field RFEH. This makes it ideal for RFID, wireless charging, and medical implants [6].

And it has two methods of energy transfer:

- **Inductive coupling:** It enables energy transfer between closely spaced coils, is effective at kilohertz and megahertz frequencies, and is widely used in wireless phone chargers and RFID tags.

- **Magnetic resonance coupling:** It allows higher-efficiency and longer-range transmission than inductive coupling, enabling multi-device charging on a single frequency with configurations like SIMO (Single-Input Multiple-Output), MISO (Multiple-Input Single-Output), and SISO (Single-Input Single-Output) [7].

4. Rectenna system in RF energy harvester architecture

The architecture of an RF energy harvester includes the following components: rectenna system, power management unit (PMU), and the load device. The term "rectenna" is derived from the combination of a receiving antenna and a rectifying circuit, which together convert ambient RF signals into usable DC power. The rectenna system plays a vital role in RFEH.

4.1 Components of rectanna system

The rectenna system includes the following components: antenna, RF filter, impedance-matching network (IMN), and rectifier.

4.1.1 Antenna

The antenna captures ambient RF signals from various sources, such as television towers, wireless routers, FM/AM radio stations, cellular base stations, and even dedicated RF sources in WPT systems. **Table 2** presents the available power density of different frequency bands [2].

Band name	Frequency (MHz)	Distance to base station (m)	Transmission power (W)	Power density ($\mu W/cm^2$)	Applications
AM Station	0.540	5,000 to 10,000	50 k	From 0.0159 to 0.0360	Radio
GSM-900	950	25 to 100	1–2	0.11 to 0.01	Phone (2G)
GSM-1800	1,747.5	25 to 100	1–2	0.1 to 0.01	Phone (2G)
UMTS-2100	2,110	11.5 to 150	1–2	0.22 to 0.02	Phone (3G)
WLAN	2,400	7 and 12	0.01/0.1/0.100	0.0001 and 0.00001	WLAN

Table 2.
Power density of different frequency bands.

4.1.2 RF filter

An RF input filter is placed between the antenna and the rectifier to suppress harmonics and eliminate unwanted frequencies, ensuring that only the fundamental operating frequency is delivered to the rectifier [5].

4.1.3 Impedance-matching network

The IMN ensures maximum power transfer by matching the impedance of the rectifier to that of the antenna. According to the maximum power transfer theorem, the load impedance should be the complex conjugate of the source impedance [3].

The IMN typically consists of a combination of resistive and reactive components (inductors and capacitors). Using purely resistive elements can result in significant power loss, so a suitable design should be chosen [4]. A properly tuned matching network can reduce transmission losses, enhance voltage gain, and improve system sensitivity.

4.1.4 Rectifier

The rectifier converts the captured RF signal into DC voltage due to the nonlinearity behavior. Typical nonlinear devices include diodes and transistors [1].

To ensure efficient rectification, a diode with the following characteristics is preferred: Low series resistance, low turn-on voltage, low junction capacitance, and high breakdown voltage. Schottky diodes are often used in RF rectifiers because of their fast-switching speeds and low threshold voltage. It's one of the diodes used in RFEH. Similarly, many rectifier topologies, as listed below, are used in rectifier design for different requirements.

- Single-diode (half-wave): Best for low-input power levels
- Full-wave bridge: Effective at higher power levels.
- Voltage doubler: Suitable for moderate power levels
- Greinacher and multistage rectifiers: Used for increasing output voltage under varying input conditions [5].

4.2 Power management unit

The PMU regulates, stores, and delivers the harvested energy to the load. Energy can be stored in micro-batteries, supercapacitors, or storage capacitors for future use [8, 9]. The PMU may also include voltage regulators, DC–DC converters, and switching circuits to optimize delivery to the end application.

4.3 Load devices

The load devices can be any sensors or devices that utilize the energy from the PMU. These may include IoT modules, medical implants, environmental monitors, or low-power communication units.

Feature	RFEH	WPT
Power source	Ambient RF sources (e.g., Wi–Fi, TV/radio stations, mobile towers, wireless routers)	Dedicated RF power transmitting antenna
Source characteristics	Unknown source: direction, polarization, frequency, and power are unpredictable	Known source: direction, polarization, frequency, and transmitting power are defined
Antenna requirements	Omnidirectional, multiband/wideband antennas for capturing random signals and polarizations	High-gain directional antennas optimized for known transmission parameters
Efficiency	Generally low, especially in low-power-density environments	Higher efficiency due to controlled and focused energy transmission
Application focus	Energy scavenging for autonomous IoT, low-power devices, and wearables	Intentional energy delivery for targeted devices, including drones, implants, and so on

Table 3.
RFEH versus WPT.

5. RFEH versus WPT

RFEH and WPT both involve transferring energy wirelessly and use the Rectenna System, but they differ in source control and application. **Table 3** explains the overall differences between them [4].

6. Antenna design requirements

The RFEH efficiency relies primarily on the design and performance of the receiving antenna. There are several parameters to consider in order to optimize maximum power capture, including the type of band (single band, multiband, or wideband), polarization, gain, directivity, beamforming, misalignment between components, sensitivity, nonlinear behavior of rectifying circuits, and impedance matching.

6.1 Multiband antenna

The receiving antenna can be single-band, multiband, or broadband. However, multiband antennas are preferred in RFEH due to their ability to harvest more energy from multiple ambient sources than single-band antennas. The total available power from a multiband antenna is the sum of the powers received at each band.

$$P_T = \sum_{i=1}^{n} P_{fi} \tag{1}$$

where:

P_{fi} = power at the i-th frequency; n = number of frequency bands; P_T = total received power [4].

Other than operating frequency and bandwidth, radiation pattern and beamwidth also play a major role. For ambient energy harvesting, omnidirectional antennas are preferred because the orientation of the incoming signal is unknown. In contrast, unidirectional antennas are suited for dedicated WPT systems because the transmitter direction is known [4].

6.2 Polarization

Polarization is another parameter that is an important concern. It refers to the orientation of the electric field vector of the EM wave [3]. A mismatch in this between the transmitter and receiver leads to significant power loss. In the case of RFEH, it is better to opt for a circularly polarized antenna as it receives energy from waves of random polarization. Using dual-linearly polarized antennas helps minimize mismatch and maximize received power. In the case of WPT systems, polarization is known and controlled, allowing for designs using linear, LHCP (left-hand circular polarization), or RHCP (right-hand circular polarization) [4]. In the study by Lavanya et al. [10], the antenna is designed to switch between bands and polarization (LHCP and RHCP). Achieving wideband circular polarization in a compact structure is challenging but beneficial for harvesting energy from diverse ambient signals [3]. **Table 4** provides a quick overview of the pros and cons of linear and circular polarization [4].

6.3 Gain and directivity

The antenna gain required for RFEH is high to improve received power, but high directivity is not always required. If the antenna is omnidirectional, moderate gain is sufficient for most applications. If the rectenna is used in WPT applications, high-gain directional antennas are used to maximize power reception from the beam [4].

6.4 Beamforming

Beamforming uses antenna arrays to form multiple high-gain beams aimed in specific directions. Beamforming overcomes orientation and alignment challenges in

Features	Circularly polarized antenna	Linearly polarized antenna
Alignment to the Tx antenna	Not necessary	Necessary
Long range power transmission/reception	No	Yes
Collection of RF energy from random polarization	Yes	No
Multipath effects	Insensitive	Sensitive
Variation of output voltage in rectenna due to change in Tx/Rx rotation	No	Yes
Optimised global efficiency	More	Less
Feasibility of energy transmission to mobile devices	More	Less
Fade resistant	More	Less

Table 4.
CP versus LP antenna in RFEH and WPT.

RFEH and WPT, making it suitable for real-time adaptive power delivery. It additionally offers dynamic beam steering to locate and track receiver positions in WPT systems, improved directional energy delivery, and higher total harvested power by combining inputs constructively using phase control, power combiners, and phase shifters [11].

6.5 Sensitivity

A major limitation of designing a rectenna is ensuring its capability to capture extremely low power in ambient conditions. To improve the rectenna's sensitivity and enable operation in ultra-low power environments, losses from the rectification device, dielectric substrate, and matching network should be limited as much as possible [4].

6.6 Matching network

An IMN connects the antenna to the rectifier and ensures maximum power transfer. It matches the antenna impedance to the complex conjugate of the rectifier input, acting as a low-pass filter (LPF) to block higher-order harmonics from reflecting back to the antenna. This reduces insertion losses and increases overall power conversion efficiency (PCE) [3]. Proper matching is vital to avoid reflections that could severely degrade system performance. A broadband impedance matching circuit can also be used for rectenna [12].

6.7 Diode in rectifier design

The diode is the core element of the rectifier, which converts RF signals to DC voltage. Energy harvesting requires zero-bias operation in a low-power range. A diode with low series resistance, low junction capacitance, large breakdown voltage, and low turn-on voltage is preferable [1]. Schottky diodes are most suitable because of their fast switching and low forward voltage. Commonly used models include the SMS7630 series, HSMS-28xx, MA4E1317 [5]. These diodes are often modeled in RF simulation tools for precise design [1]. **Figure 2** shows a similar Schottky diode model.

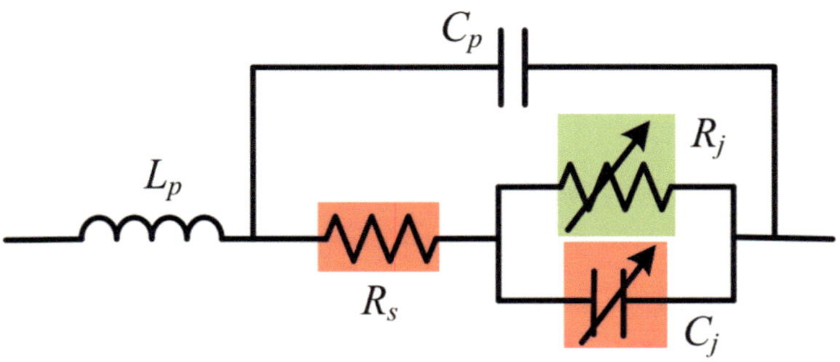

Figure 2.
Equivalent circuit model of the Schottky diode [1].

7. Metamaterial and metasurface

The terms metamaterials and metasurfaces refer to artificially engineered structures designed to exhibit EM properties that are not found in nature. These materials are composed of subwavelength-scale unit cells – often referred to as artificial "atoms" or "molecules" – fabricated from metallic or dielectric components. By tailoring their shape, size, arrangement, and interaction, it is possible to achieve tunable permittivity and permeability, including values that are positive, near-zero, or even negative [13].

These engineered materials play a significant role in WPT and wireless energy harvesting (WEH). The role of metamaterials and metasurfaces is to enhance antenna gain and miniaturization. One common method involves using metamaterial superstrates, which are layered above a standard patch antenna to enhance its gain, directivity, and efficiency while maintaining a compact size [4]. Metamaterials also have an influence on energy harvesting devices [14].

7.1 Metamaterial and metasurface in rectenna design

Wireless charging technologies that eliminate the physical cables used in mobile phones and electric vehicles rely on WPT. Though near-field coupling is currently the most commonly used method, there is increasing interest in far-field WPT due to its potential for longer-distance applications. One of the challenges in both WPT and WEH is the low efficiency of converting RF energy into DC. Enhancing the conversion efficiency remains a key research focus. Metamaterials and metasurfaces have shown great promise in addressing these issues by significantly improving power transfer efficiency and extending operational range – making them highly beneficial in rectenna design for RFEH.

8. Applications

Ambient RFEH and WPT are evolving as sustainable, cost-effective, and environmentally friendly alternatives for powering low-power electronic devices. These technologies are particularly suitable for a variety of applications across sectors such as mining, agriculture, manufacturing, healthcare, and smart cities. A widely adopted application of WPT is in passive ultra-high frequency RF identification (UHF RFID) systems, commonly operating around 900 MHz. In such systems, passive RFID tags are wirelessly powered during communication with an RFID reader, eliminating the need for embedded batteries. Additionally, RFID tags and sensors embedded in building materials and tiles can enable novel and energy-efficient implementations in IoT and WPT-driven architectures [4]. The subsequent sections will delve into the practical applications of RFEH and WPT in key areas such as IoT, medical devices, and wireless communication. Wearable circularly polarized [15] and ultra-wideband antennas are designed for 5G and healthcare[16].

8.1 IoT applications

In the context of IoT, RF energy scavenging and WPT emerge as better solutions, offering a sustainable, long-term power source that can enable the continuous operation of IoT devices in remote or inaccessible environments. Other mechanisms, like wireless information transfer (WIT) and WET, are also discussed further.

To further enhance efficiency, researchers have explored three main techniques for combining energy from rectenna arrays:

1. RF power combining – RF signals from individual rectennas are first combined and then rectified.

2. DC power combining – Each rectenna rectifies the RF signal independently, and the resulting DC outputs are then combined.

3. Hybrid power combining – A mix of RF and DC combining designed to leverage the strengths of both approaches [17].

These techniques aim to optimize the total harvested energy, improve system reliability, and extend the operational lifespan of autonomous IoT systems.

8.1.1 Compact multiband RF energy harvester

A compact multiband RF energy harvester operates across five key frequency bands: 0.94 GHz, 1.8 GHz, 2.1 GHz, 2.6 GHz, and 3.5 GHz. This system integrates three main components: an omnidirectional modified monopole antenna, a dual-branch multiband rectifier, and a power management module. To maintain omnidirectional radiation while ensuring a compact form factor, the current distribution in the antenna suppresses high-order modes common in conventional monopole designs. The resulting antenna has a compact footprint of just $0.24\lambda L \times 0.22\lambda L$, where λL corresponds to the wavelength at 0.94 GHz. The system's PCE under an input power of – 20 dBm is noteworthy: 19.2% at 0.94 GHz, 24% at 1.8 GHz, 24.5% at 2.1 GHz, 17.7% at 2.6 GHz, and 16.4% at 3.5 GHz. A dual-branch rectifier design is implemented (see **Figure 3**), where one branch handles 1.8 GHz and 3.5 GHz, and the second branch covers 0.94 GHz, 2.1 GHz, and 2.6 GHz. In outdoor ambient environments, the rectenna delivers a peak DC output power of 18.7 μW, which is ample to power low-power IoT devices such as temperature sensors and environmental monitors [18]. The proposed rectenna was integrated with a power management and storage system, which includes the BQ25570 energy management module and the BMT005R5L224FA supercapacitor.

8.1.2 WIT and WET

RFEH with energy-aware data collection offers an effective solution to extend the operational life of IoT devices while minimizing their dependency on batteries. The core concept involves reducing the amount of data collected by IoT devices using sparse data sampling and then reconstructing the complete dataset via matrix completion techniques. These techniques leverage the spatial–temporal correlation within the sensed data, enabling the devices to stay in low-power dormant states for longer durations and wake only intermittently for sampling. This significantly

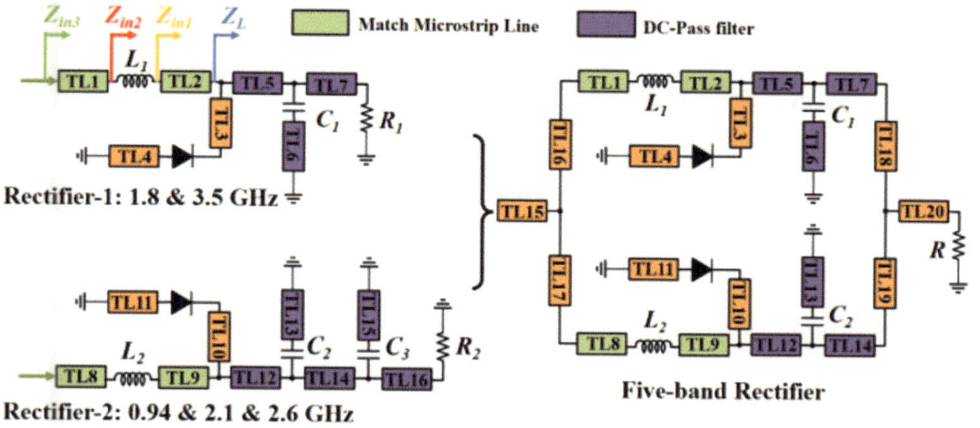

Figure 3.
Dual-branch rectifier design.

lowers energy consumption while maintaining data integrity with minimal information loss.

To support this energy-saving strategy, an adaptive RF energy management mechanism is introduced. This mechanism operates within a Data and Energy Integrated Network (DEIN), where both WET and WIT are dynamically managed between the DEIN Gateway (DEING) and the DEIN Nodes (DEINN).

In this system:

The DEINN and DEING autonomously switch between WET and WIT modes based on energy requirements and data transmission needs.

This dynamic adaptation ensures optimal use of harvested RF energy, maximizing energy efficiency while minimizing idle energy waste.

The integration of energy harvesting with adaptive data acquisition and transmission allows the batteries of IoT devices to last significantly longer or even operate perpetually in some cases.

Simulation results validate the proposed integrated system, demonstrating enhanced efficiency in both data transmission and energy utilization [19].

8.1.3 Uniform coverage rectenna for IoT applications

A Space-Diverse Rectenna System (SDRS) (see **Figure 4**) has been developed to overcome the limited power harvesting and sensitivity to the orientation and positioning of IoT sensor nodes relative to the transition in conventional rectennas. The maximum RF-to-DC conversion efficiency is 64% (simulated) and 60.5% (measured) for two-series EndFire Rectifying Antennas (EFRA) elements.

The SDRS features full-wave rectification, capturing both positive and negative RF waves for improved harvesting. Its mirror symmetry ensures orientation-insensitive reception in both azimuth and elevation planes and has the ability to harvest an open-circuit voltage of 1,600 mV, which is sufficient for powering low-power IoT nodes.

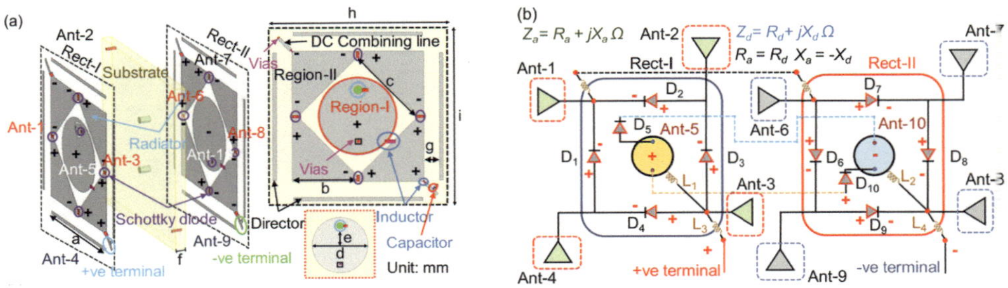

Figure 4.
(a) Layout of the proposed SDRS and (b) system diagram.

Unlike conventional systems, the SDRS:

- Requires no beamforming circuits or external matching networks, making it compact and ideal for miniaturized IoT nodes.
- Employs direct conjugate matching between the antenna and rectifier to minimize insertion loss.
- Provides uniform and isotropic RF energy coverage, suitable for deployment in RF-constrained smart environments.

The SDRS has been validated through experimentation and shown to support various IoT applications, including electromyogram (EMG) sensing, thermocouple readings, and gas detection, operating effectively at input power levels as low as -12 dBm.

This makes SDRS a promising solution for battery-less IoT operation and uniform WPT across diverse application environments [20].

8.2 Medical application

Rectennas are crucial in various medical applications, especially for powering wearable health sensors and implantable biomedical devices. They are also used in monitoring cardiac conditions and can be integrated into systems for deep brain stimulation and neural activity recording. Their ability to wirelessly harvest energy makes them ideal for compact, low-power medical technologies that require reliable and continuous operation without frequent battery replacements. The applications of rectennas in medical technologies are briefly discussed below.

8.2.1 Flexible meta-patch rectenna array

A textile-based meta-patch antenna was developed using a felt fabric substrate, incorporating a grounded metasurface composed of a 3 × 3 unit cell array instead of a conventional ground plane. A meta patch antenna incorporates an SRR (Split Ring Resonator) metasurface. The SRR metasurface unit cell is designed to achieve precise control over radiation direction. To operate the rectenna effectively, even under low input power conditions, a seven-stage Cockcroft–Walton voltage multiplier (CWVM) rectifier is used. This high sensitivity to weak RF signals makes the

Wearable medical sensors	Power consumption (μW)
Temperature sensors	113 pW and 1.4 μW
Cardiac activity sensors	0.3 μW
Electrocardiogram (ECG) amplifier sensors	2.76 μW
Electronic-nose sensors	250 μW

Table 5.
Wearable medical sensors and their respective power consumption.

rectenna highly suitable for low-power, body-worn medical devices, where efficient WEH is critical. This design delivers an efficiency of 77%, a high gain of 7 dB, and a wide bandwidth of 120 MHz. The performance of the 2 × 2 meta-patch rectenna array was evaluated under typical Wi–Fi conditions by measuring its DC output voltage and power at various distances from a router. At just 10 cm from the router, the array produced a DC output power of 102.1 μW [21]. A few examples of WMS and their respective power consumption are provided in **Table 5** [21].

8.2.2 Rectenna for biomedical implantable devices

RFEH offers a promising solution in the field of biomedical implants for wirelessly powering miniaturized internal devices. The use of on-chip hybrid rectifiers is a notable advancement in biomedical implants. It operates at 2.4 GHz, a frequency that supports both compact design and high energy density. By using modern CMOS technology, these rectifiers are well-suited for powering implantable medical applications. Some key components and features include an on-chip IMN, adaptive rectification path, high-frequency operation, efficiency, compact and single-ended architecture. This rectenna architecture demonstrates strong potential for powering energy-efficient biomedical implants without the need for battery replacement or wired charging, thereby enhancing device reliability and patient comfort [22].

8.2.3 Coil rectenna array

A coil rectenna array has been proposed to overcome the power loss due to misalignment between Tx and Rx coils in WPT systems by integrating both rectifying and receiving components into a compact structure using multi-layer Printed Circuit Board (PCB) technology. The system offers a single DC output port and incorporates all necessary circuitry directly within the antenna structure. This integration leads to a cost-effective, space-efficient, and reliable solution for the wireless powering of miniaturized biomedical and wearable devices.

The rectenna array is specifically designed to maintain a consistent output voltage (V_o) across the Rx region, improving tolerance to lateral misalignment. The design includes two orthogonal receiver coils in addition to a standard planar spiral coil. This configuration effectively captures all three orthogonal components of the magnetic field (H-field), thereby enhancing power reception regardless of positional deviation.

A DC combining technique is employed to merge the output from each coil efficiently, delivering a stable power supply to the load. Optimization of the receiver coil parameters has been shown to further enhance performance compared to traditional Rx antenna configurations. Experimental results confirm high system efficiency, with a rectification efficiency of 53.30% and a DC combining efficiency of 93.65% [23].

8.2.4 Dual-port implantable rectenna

A compact and highly integrated dual-port implantable rectenna network (see **Figure 5**), designed specifically for multisite WPT in miniature vascular implants. The system enables Simultaneous Wireless Information and Power Transfer (SWIPT), utilizing a flexible transmitter dipole and dual-band implantable antennas. Each rectenna, measuring the compact dimensions of 7.0 × 7.0 × 0.635 mm^3, operates at 915 MHz and 2.45 GHz. It offers excellent isolation greater than 25 dB between the two ports. Fabricated on a single PCB, this low-profile rectenna is optimized for implantable biomedical systems, providing a space-efficient and seamless solution for battery-less vascular implants. The design successfully demonstrates wireless charging through an integrated rectifier and energy storage module. Under experimental conditions, the rectenna achieved a maximum charging efficiency of 48.97% at an input power of 8 dBm, marking one of the highest recorded efficiencies for such devices. Notably, this is the first demonstration of a multisite implantable rectenna array (see **Figure 5**), and the study evaluates the energy levels harvested over time using a flexible wearable transmitter [24].

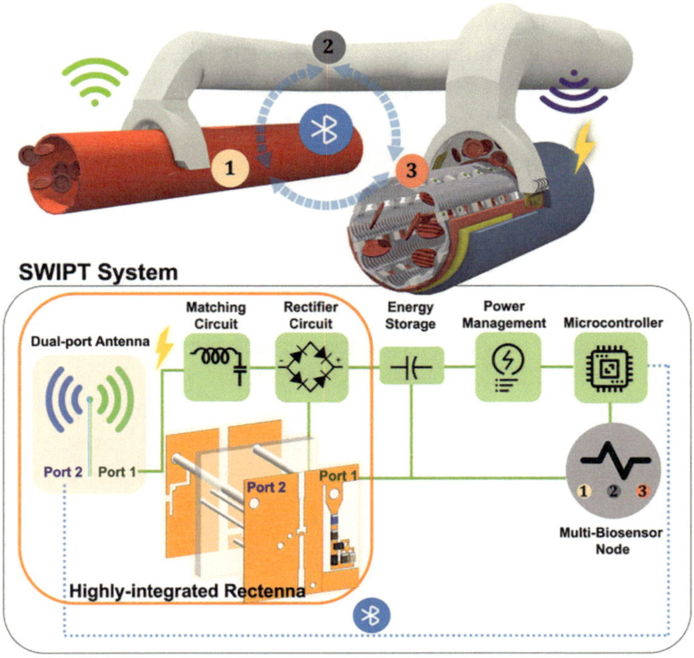

Figure 5.
Two-port rectenna for SWIPT.

8.2.5 Radiating near-field patch rectenna

A radiating near-field RF energy harvester consists of both the rectenna and its transmitter. In this setup, the transmitter is positioned outside the body, close to the skin. While transmitters are often realized as coils, here a patch array is used instead to create the desired radiating near-field. Conventional near-field charging relies on magnetic coupling between coils, but such systems are highly vulnerable to misalignment between the transmitter and receiver. By contrast, the design in the study by DeLong et al. [25] applies WPT principles with radiating antennas, enabling reliable power delivery even with small receiving antennas (27.5 mm × 19.75 mm) and significantly reducing sensitivity to misalignment. Operating at 2.4 GHz, the system achieved an energy conversion efficiency of up to 47.7%. Positioned in the radiated near-field region, the transmitting antenna successfully delivered approximately 1.2 mW of power over a 42 cm distance to the rectenna. The design was also tested for misalignment tolerance, showing a maximum power drop of 7.5 dB with a lateral displacement of 15 cm. Compared to traditional coil-based systems, this approach provides a more stable and less position-sensitive solution for WPT, particularly suitable for medical applications [25].

8.3 Applications in communication

8.3.1 Decoupling slots for enhanced received power in far-field WPT

The rectenna array system incorporates decoupling slots to improve isolation between antenna elements. It enhances received power regardless of the polarization orientation between the transmitting (Tx) and receiving (Rx) antennas. It also enables miniaturization, making it ideal for compact, far-field WPT systems. The received power is 50% higher compared to conventional rectenna arrays when Tx and Rx polarizations are aligned. If a 90° polarization mismatch occurs, the proposed array continues to receive 1.2 mW, while traditional rectenna arrays fail to receive any usable power. It ensures the system's robustness against polarization misalignment [26].

8.3.2 Self-sustained RFEH rectenna for sensor node applications

RFEH is gaining recognition as a promising approach for powering wireless and ubiquitous IoT devices. However, key challenges remain – primarily the limited ambient RF power density, the sensitivity of the receiver, and the effectiveness of the antenna design. This study experimentally investigates these challenges, particularly focusing on the difficulty of harnessing sufficient energy from ambient RF sources to support a fully functional RFEH system.

To address these limitations [2], this study explores various antenna design trade-offs involving size, frequency, and gain, and outlines practical design guidelines. A proof-of-concept is demonstrated using a dedicated RF source to successfully power a Texas Instruments CC1350 SoC-based IoT sensor node. The system's performance was evaluated through experimental measurements of temperature, power generation, and energy dissipation.

The results reveal that the proposed RFEH setup is capable of powering and sustaining more than one sensor node. After charging a supercapacitor to 3.3 V, the system delivers approximately 2 mW of power. This is sufficient to maintain the operation of the IoT-enabled sensor node, especially when temperature data is recorded at one-minute intervals – allowing enough time for the supercapacitor voltage to be replenished between measurements [2].

8.3.3 Fully woven textile rectenna

Flexible, textile-based rectennas are emerging as feasible solutions for WPT in wearable and communication systems. A fully woven textile rectenna operating in the 2.4 GHz ISM band was designed specifically for WPT applications (see **Figure 6**). The design consists of two conductive layers woven with silver-coated polyamide ELITEX 117/f17 two-ply threads, and separated by a dielectric layer woven with polyester threads. All the layers are held together with polyester binder threads, which are woven at the same time as the rest of the structure, providing an indivisible multilayer fabric piece with improved robustness compared to other textile nonwoven approaches. The antenna consists of a square patch excited via an offset T-match network, which is directly linked to the rectifier. This feeding approach enables effective impedance matching with the rectifier without the need for additional matching components, resulting in a compact and integrated design. The placement of the feed at one corner of the patch is intentionally used to achieve circular polarization, enhancing the antenna's reception capabilities. The rectifier circuit employs a single Schottky diode, chosen for its simplicity and ease of integration into textile substrates. Measured results show an RF-to-DC conversion efficiency of approximately 50% at an input power level of – 10 dBm, with a peak DC output power of around 50 μW achieved near 2.4 GHz – indicating optimal impedance matching at that frequency. The system was tested at various distances (0.3, 0.65, and 0.8 m), corresponding to estimated available RF input powers of – 6, – 12, and – 14 dBm, respectively, assuming 0 dB antenna gain [27].

8.3.4 Flat panel rectenna (FPR)

A flat panel rectenna (FPR) is proposed to overcome the spatial limitations commonly associated with omnidirectional antenna-based rectennas for microwave-based wireless power-charged sensor applications. The design enables seamless integration into self-powered systems while minimizing uncovered or inactive zones. It optimizes both space utilization and energy capture, thereby ensuring more reliable and continuous operation of wirelessly powered devices across diverse applications. Conventional rectenna systems (as depicted in **Figure 7(a)**) suffer from limited directional coverage, resulting in intermittent wireless power charging (WPC) performance due to orientation sensitivity. In contrast, the proposed FPR (illustrated in **Figure 7(b)**) offers enhanced orientation flexibility and more consistent energy harvesting performance. This extended beam coverage makes the FPR a robust candidate for practical, self-powered sensor solutions, as demonstrated through its integration with the low-power heart rate sensor [28].

8.3.5 Wireless sensor networks

Wireless sensor networks (WSNs) consist of multiple sensor nodes that work collaboratively to sense environmental data, perform basic computations, and

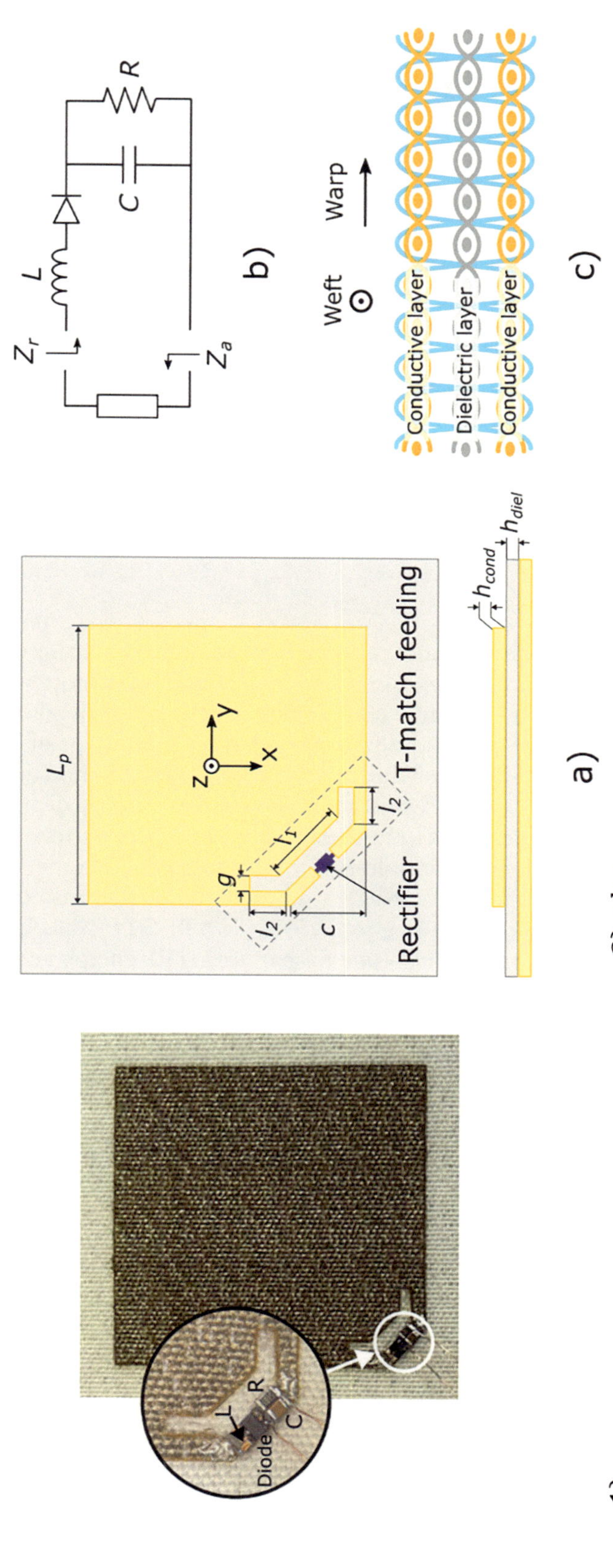

Figure 6.
Fully woven textile antenna. (a) Top and front view, (b) topology of the single Schottky diode-based rectifier, and (c) multilayer structure.

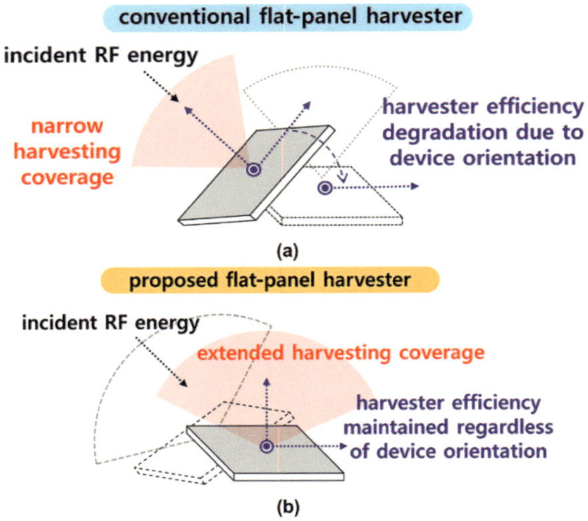

Figure 7.
Flat panel RFEH with (a) conventional patch-based antenna coverage and (b) proposed rectenna coverage.

communicate with neighboring nodes to share information. The major problem that WSNs face is the limited power availability for sensor nodes, as most sensor nodes are battery-powered and are typically constrained in processing capability, data storage, and energy capacity. In addition to the inherent limitations of battery life, issues such as leakage current – even during idle states – further drain energy. Harsh environmental conditions, such as extreme temperatures, can also degrade battery performance and contribute to environmental pollution due to battery disposal. Researchers are actively investigating alternative and more reliable energy sources to extend the operational lifespan of WSNs. To address these limitations, RFEH is being explored as a sustainable power source for WSNs. Two RFEH models have been proposed: one utilizes separate radios – one for harvesting RF energy and another for communication – while the other employs a single radio for both data transmission and energy harvesting simultaneously [4].

9. SWIPT technology

SWIPT technology offers the capability to integrate both energy harvesting and data transmission in modern wireless sensor networks. This concept of transmitting energy and information simultaneously was first introduced by Varshney [29].

In the study by Zhou et al. [29] (see **Figure 8**), the received RF-band signal $y(t)$ is initially transformed into a complex baseband signal $y\_b(t)$. It is then sampled and digitized using an analog-to-digital converter (ADC) for data decoding. In the rectenna, the same received RF signal $y(t)$ is converted into a DC signal $i\_DC(t)$ through a rectifier, typically consisting of a Schottky diode and a passive LPF. This DC signal is used to charge a battery for energy storage. Given an input voltage proportional to $y(t)$, the output current $i(t)$ of a Schottky diode can be expressed as:

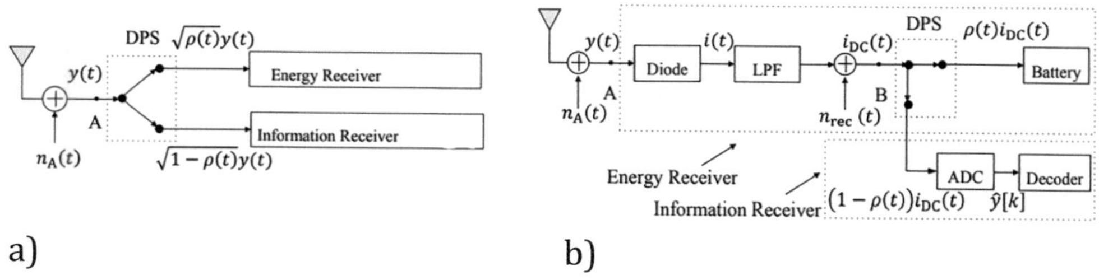

Figure 8.
(a) Architecture for the separated information and energy receiver and (b) architecture for the information and energy receiver.

$$i(t) = I_s\left(e^{\gamma y(t)} - 1\right) \tag{2}$$

where γ is the reciprocal of the thermal voltage of the Schottky diode, and I_s is the saturation current. Assuming the battery's stored energy is linearly related to $i_{DC}(t)$, with a conversion efficiency $0 < \zeta < 1$, it is further assumed that the contribution of noise (from both the antenna and rectifier) to the harvested energy is negligible and can be treated as a small constant. Therefore, the harvested energy stored in the battery, denoted as Q (in joules), for a symbol period of one, is expressed as:

$$Q = \varsigma E[i_{DC}(t)] = \varsigma hP \tag{3}$$

where:
ζ is the conversion efficiency,
h is the channel power gain,
P is the transmit power.
Energy harvested due to noise (antenna and rectifier noise) is considered negligible and is ignored in the calculation [29].

10. Challenges and conclusion

The chapter discussed the introduction to the rectenna system and its applications in IoT, medical, and communication fields. It also provided an introduction to a few terms like SWIPT, WET, and WIT. Significant challenges faced by RFEH systems are primarily due to the low levels of ambient RF power available in typical environments. The amount of energy that can be scavenged depends on several factors, including the characteristics of the received signal – such as polarization, bandwidth, frequency, and power flux density – as well as telecommunication traffic density and antenna orientation. Since these conditions are often unpredictable, RFEH antennas must be capable of capturing signals with varying polarization states and bandwidths. A major practical hurdle is generating sufficient power to activate the rectifier circuit, especially when relying on a single antenna [4]. Additional issues include impedance mismatch losses in rectifiers, sensitivity to antenna orientation, and polarization mismatches. These obstacles can be mitigated using techniques such as reconfigurable antennas, beamforming,

MIMO (Multiple Input Multiple Output) configurations, dynamic impedance matching, and energy-aware communication protocols for IoT devices.

Despite these challenges, RFEH holds great promise for enabling battery-free and low-maintenance wireless sensor networks, particularly in IoT and Industry 4.0 applications. In environments like shopping malls, coffee shops, and airports, where wireless devices are prevalent, the EM energy emitted by these devices can be harvested and converted into electrical energy to power small devices [30].

Author details

V. Lavanya[1*], C. Rimmya[2], S. Piramasubramanian[1] and Madhan M. Ganesh[1]

1 Madras Institute of Technology, Chennai, India

2 College of Engineering Guindy, Chennai, India

*Address all correspondence to: vlavanya0903@gmail.com.

References

[1] Niotaki K, et al. RF energy harvesting and wireless power transfer for energy autonomous wireless devices and RFIDs. IEEE Journal of Microwaves. 2023;**3**:763–782

[2] Raghav KS, Paliwal R, Singhal N, Kumar MS, Bansal D. Design and development of self-sustained RF energy harvested rectenna for sensor node application. IEEE Transactions on Instrumentation and Measurement. 2025;**74**:1–12

[3] P L, Mohanta HC, "Harvesting RF electromagnetic energy for smart sensors in smart communication systems- An overview," 2024 2nd International Conference on Signal Processing, Communication, Power and Embedded System (SCOPES). Paralakhemundi Campus, Centurion University of Technology and Management, Odisha., India, 2024

[4] Ullah MA, Keshavarz R, Abolhasan M, Lipman J, Esselle KP, Shariati N. A review on antenna technologies for ambient RF energy harvesting and wireless power transfer: Designs, challenges and applications. IEEE Access. 2022;**10**:17231–17267

[5] Surender D, Khan T, Talukdar FA, Antar YMM. Rectenna design and development strategies for wireless applications: A review. IEEE Antennas and Propagation Magazine. 2022;**64**(5):16–29

[6] Garnica J, Chinga RA, Lin J. Wireless power transmission: From far field to near field. Proceedings of the IEEE. 2013;**101**(6):1321–1331

[7] Kumawat Y, Shukla S, Verma D, Rathore PS, "Wireless Energy Harvesting and Transfer: A Comprehensive Review of Recent Developments," 2023 IEEE Renewable Energy and Sustainable E-Mobility Conference (RESEM). Bhopal, India, 2023

[8] Kar PC, Islam MA. "Design and performance analysis of a rectenna system for charging a mobile phone from ambient EM waves,". Heliyon. 2023;**9**(3):e13964

[9] Surender D, Khan T, Talukdar FA, De A, Antar YMM, Freundorfer AP. Key components of rectenna system: A comprehensive survey. IETE Journal of Research. 2020;**68**(5):3379–3405

[10] Lavanya V, Rimmya C, Kanisha RKR, Madhan MG. A novel miniaturized metamaterial inspired frequency reconfigurable antenna with switchable polarization for sub 6 GHz band. Physica Scripta. 2024;**99**(12):125501

[11] Abdollahi S, Bakhtafrouz A, Maddahali M. Design of a beamforming Tri-Band Rectifier for RF energy harvesting. IEEE Access. 2025;**13**:34802–34811. DOI: 10.1109/ACCESS.2025.3544146

[12] Rabin Kanisha RK, Piramasubramanian S, Madhan MG. Design and analysis of broadband impedance matching network for rectenna. Journal of the Instrument Society of India. 2022;**52**(1-4):89–97

[13] Huang Y, Zhou J, Li L, "Wireless power transfer and energy harvesting using metamaterials and metasurfaces," 2022 International Workshop on

Antenna Technology (iWAT), Dublin, Ireland, 2022

[14] Góra P, Łopato P. Metamaterials' application in sustainable technologies and an introduction to their influence on energy harvesting devices. Applied Sciences. 2023;**13**(13):7742

[15] Sabban A. Wearable circular polarized antennas for health care, 5G, energy harvesting, and IoT systems. Electronics. 2022;**11**(3):427

[16] Sabban A. Ultra-Wideband systems with energy harvesting units for sensors, 5G, IoT and medical systems. Journal of Sensor Technology. 2021;**11**(1):1–17

[17] Kim S, Dong HJ, Lee HL. Hybrid power combining rectenna array for wide spatial coverage of self-powered IoT devices. IEEE Internet of Things Journal. 2024;**11**(22):37302–37311

[18] Wang W, Zhang J, Sun X, Zu H, He D. A compact five-band omnidirectional RF ambient energy harvesting system for IoT applications. IEEE Access. 2025;**13**:86110–86119

[19] Wang Y, Yang K, Wan W, Zhang Y, Liu Q. Energy-Efficient data and energy integrated management strategy for IoT devices based on RF energy harvesting. IEEE Internet of Things Journal. 2021;**8**(17):13640–13651

[20] Kumar M, Sharma A. A compact planar multisector rectenna array with full-wave rectification for 3-D uniform wireless powering of IoT nodes. IEEE Transactions on Antennas and Propagation. 2025;**73**(6):3470–3480

[21] Alkhalaf HY, Ahmad MY, Ramiah H, Hossain AKMZ, Azam SMK, Thiha A. Flexible Meta-Patch rectenna array for energizing low-power wearable medical sensors. IEEE Access. 2024;**12**:121570–121585

[22] Nagaveni S, Hunasigidad P, Pathak D, Dutta A. On-Chip configurable RF energy harvester for biomedical implantable devices. IEEE Transactions on Circuits and Systems I: Regular Papers. 2024;**71**(11):5030–5039

[23] Srivastava VK, Sharma A. A coil rectenna array design to harvest all H-Field components for lateral misalignment tolerant wireless powering of bio-medical implant devices. IEEE Journal of Electromagnetics, RF and Microwaves in Medicine and Biology. 2024;**8**(1):59–67

[24] Zhang J, et al. Highly integrated two-port rectenna system for wirelessly powering multisite autonomous vascular implantable networks. IEEE Transactions on Microwave Theory and Techniques. 2025;**73**:3655–3666

[25] DeLong BJ, Kiourti A, Volakis JL. A radiating near-field patch rectenna for wireless power transfer to medical implants at 2.4 GHz. IEEE Journal of Electromagnetics, RF and Microwaves in Medicine and Biology. 2018;**2**(1):64–69

[26] Kim MS, Park DH, Elhefnawy M, Lee W-S. Densely placed rectenna array with decoupling slots for enhanced received power in far-field wireless power transfer. IEEE Access. 2025;**13**:43249–43258

[27] Fernández M, Vázquez C, Hoeye SV. Fully-woven textile implementation of a circularly polarized rectenna for wireless power transfer applications. IEEE Access. 2025;**13**:110336–110343

[28] Han Y, Kim E, Lee HL. Flat-Panel-Rectenna with broad RF energy harvesting coverage for wireless-powered sensor applications. IEEE Access. 2025;**13**:6146–6153

[29] Zhou X, Zhang R, Ho CK. Wireless information and power transfer: Architecture design and rate-energy tradeoff. IEEE Transactions on Communications. 2013;**61**(11): 4754–4767

[30] Agieb R, Amer A, Mansour I, Solyman A, Yahya K, Ahmed Samir. "Improving energy harvesting system from ambient RF sources in social systems with overcrowding". International Journal of Crowd Science. 2025;**9**(1):13–28

Section 2

Advanced Antennas and Modules

Chapter 4

Reconfigurable Intelligent Surfaces: Principles and Design Considerations

Karamot Kehinde Biliaminu, Jonathan Rodriguez, Felipe Gil-Castiñeira and Joaquim Bastos

Abstract

A reconfigurable intelligent surface (RIS) is a technology within wireless communications, particularly related to future networks like 6G. An RIS is a planar structure containing a large number of passive, low-cost, and electronically controllable elements, called meta-atoms or unit cells, which can change the phase, amplitude, or polarization of an incident electromagnetic wave, thus giving the ability of such a surface to reflect or scatter wireless signals in a controlled manner. The ability of RIS technology to control the wireless propagation environment in real time leads to several benefits, such as expanding the signal coverage, enhancing data rates, and mitigating interference. The various deployment scenarios for RIS can be classified as static or mobile configurations. Static RIS configurations involve the surface being affixed to supporting infrastructures (i.e., walls, ceilings, rooftops). Mobile RIS configurations entail attaching the surface to moving platforms (i.e., unmanned aerial vehicles (UAVs) or any other vehicles, robots), offering a more dynamic and flexible means for coverage in often complex environments. Another consideration for connectivity configurations is whether the connection involves either a single RIS or multiple RIS units, whereby the joint cooperation of one or more RIS units can provide improved spatial diversity and communication link reliability. This chapter offers a general understanding of RIS, including operation under different scenarios, connectivity models, and optimization techniques for RIS-assisted systems. It also discusses the irregular RIS. Furthermore, it addresses key performance metrics, such as spectral efficiency and energy efficiency, as well as the factors that impact those metrics.

Keywords: reconfigurable intelligent surface, RIS scenarios, RIS connectivity, performance metrics, irregular RIS

1. Introduction

Reconfigurable intelligent surfaces (RIS) feature N reconfigurable elements, which all operate like tiny antennas that receive signals and re-radiate the received signals [1]. RIS simplifies integration because it can simply be installed without requiring any infrastructure modifications, making it a more affordable alternative to improving a wireless communication system. These elements can change various parameters of the wireless signals to maximize the overall beamforming potential to specific receivers, improving signal quality and reducing interference. Regarded as a type of passive beamforming technology, RIS can offer strong benefits such as enhancing network and energy efficiency [2, 3]. In addition, RIS is very successful at increasing connectivity in a non- line-of-sight (NLOS) situation by eliminating physical barriers and increasing coverage for urban and indoor environments as part of the deployment. In conclusion, RIS allows network designers to incorporate greater reliability, increased transmission speeds, and better spectrum usage. It is an essential component of future generation wireless communication technologies, such as beyond 5G/6G [4–6].

In contrast to what previous wireless generations achieved, 6G intends to provide extraordinary connectivity by offering ultra-high data rates, ultra-low latency, and ultra-reliable connectivity to support applications across various sectors, with a plethora of newly offered services. An important part of 6G is artificial intelligence (AI) that will play a role in delivering network-wide autonomous [7] and optimized service delivery, self-healing, and maintenance [8, 9]. In this context, AI is foreseen to work in synergy with THz-enabled RIS to offer unique and unprecedented opportunities to passively and intelligently reconfigure the wireless environment. In October 2021, the European Telecommunications Standards Institute (ETSI) established the Industry Specification Group (ISG RIS) to acknowledge the importance of RIS [10].

Purpose of ISG RIS is to establish worldwide standards for RIS technologies and promote their acceptance in future communication systems [11].

The fundamental principle behind RIS lies in its ability to individually control the response of each of its reflecting elements. Each element can impose specific phase shifts on the incident electromagnetic wave. RIS can:

- Steer the signal: Direct the reflected signal toward a specific receiver, effectively creating a virtual line-of-sight (LoS) path or enhancing an existing one. This is akin to beamforming but performed passively by the surface itself.

- Focus the signal: Increase signal strength and coverage in specific locations by concentrating the reflected energy at a specific point.

- Cancel interference: Redirect unwanted signals or nullify interference by steering them away from the receiver or by creating destructive interference patterns.

- Enhance signal-to-noise ratio (SNR): By coherently combining reflected signals at the receiver, RIS can significantly boost the received signal power, leading to higher SNR and improved data rates.

1.1 Optimization techniques for RIS-assisted systems

Optimizing the performance of RIS-assisted wireless communication systems is crucial to fully harness its potential. The goal of RIS optimization often involves maximizing performance metrics such as achievable data rate, energy efficiency, or coverage, while minimizing interference or power consumption – which often requires the joint optimization of the active beamformers at the base station (BS) or access point (AP), and the passive phase shifts at the RIS [12]. Significant optimization methods include:

- **Alternating optimization (AO):** AO is a popular iterative algorithm that separates the optimization problem into non-coupled subproblems, which can be solved subsequently. Specifically, this often means optimizing the phase shifts at the RIS while keeping the BS precoder fixed, and then solving for the optimal BS precoder while treating the RIS phase shifts as fixed [13, 14]. The process continues until convergence is achieved.

- **Machine learning (ML) and deep learning (DL):** Given the complexities of large-scale deployments of RIS and fast-moving wireless environments, applications of ML and DL techniques are increasingly being used for RIS optimization [8]. In fact, many works have applied ML techniques to optimize RIS-assisted wireless networks [15–18]. Given these data-driven approaches, they can learn optimal phase shift configurations from historical data or by monitoring ongoing events, providing more robust and adaptive solutions. The integration of DL with reinforcement learning (RL), known as deep reinforcement learning (DRL), has enormous potential to improve the ability to effectively solve complex problems and provide optimal efficiency and adaptability [19]. For example, in the study by Huang et al. [20], DRL has successfully been applied for designing RIS reflections and BS precoding optimization problems in multiuser MISO communication systems.

- **Particle swarm optimization (PSO) and genetic algorithms (GA):** GA and PSO heuristic optimization algorithms are effective at addressing non-convex problems. A GA or PSO scheme could be used to search for near-optimal approximate solutions for RIS design problems, particularly for designing RIS phase shifts when using discrete RIS phase shifts. PSO is a metaheuristic optimization technique [21, 22] that is inspired by the behavior of groups of organisms, in which individuals adjust their movement based on both personal experience and that of fellow peers in order to efficiently explore a solution space. Each particle in PSO moves based on its own best-known position but also retains knowledge of the best position found by the overall swarm, which is updated as new solutions are discovered. This coordinated movement aids in guiding the swarm overall toward optimal solutions [23–25]. A relatively new approach to address optimization challenges in electromagnetics has been proposed using the PSO algorithm and has proven successful in solving several challenging EM design challenges [26]. GA algorithms are powerful, stochastic search techniques that emulate the process of natural selection by evolving a set of candidate solutions using mechanisms like selection, crossover, and mutation [27, 28]. GA is well-suited for RIS since it can explore the solution space quickly and also deal with the complexity and nonlinear nature of the problem [27].

- **Channel estimation and feedback:** To optimize the performance of RIS-assisted networks, it is necessary to have precise channel state information

(CSI) [29]. Performance gains with RISs are often dependent on accurate CSI [30]. Channel estimation is an important optimization process that allows the network to estimate channel characteristics between the BS, RIS, and end users. When the RIS has accurate CSI, it can reconfigure its phase shifts to improve beamforming, spectral efficiency (SE), and communication reliability. Inconsistent CSI will impair the RIS's ability to manipulate signal propagation toward the direction of the end users, resulting in a drastic reduction in system performance.

2. RIS and connectivity scenarios

This section elaborates on the potential RIS and connectivity scenarios.

2.1 RIS scenarios

An RIS scenario deployment can either be static, where all the network nodes (i.e., BS, RIS, and user equipment [UE]) are fixed, or dynamic, where the components (RIS and the UE(s)) could be mobile.

- **Static RIS and static UE**
  A static RIS and static UE arrangement is a scenario that allows a UE and an RIS to communicate wirelessly, such that the RIS and the UE are both stationary. The RIS is basically placed on a fixed surface (i.e., wall of a building), and the UE does not move when engaging in wireless communication via the RIS. Static arrangements provide low-complexity, energy-efficient communications and would be beneficial for scenarios where only indoor coverage is possible or in fixed wireless AP regions.

- **Mobile RIS and static UE**
  The mobile RIS and static UE scenario is a communication context where the RIS can be mounted on a mobile platform such as a drone, vehicle, or robot, while the UE remains stationary. By having a mobile RIS, the RIS can, in real-time, move and orient itself in order to optimally reflect the signals toward the static UE from different moving positions. The mobile RIS allows for increased flexibility in mitigating obstructions, enhancing coverage, and improving communication performance in dynamic and challenging locations. However, considering a mobile RIS means the communication system needs to handle new challenges, such as real-time beam alignment, rapid CSI, and low-latency control of the RIS to guarantee expected performance.

- **Static RIS and mobile UE**
  Static RIS and Mobile UE setups refer to a situational mode of wireless communication where the RIS is located at a fixed position (i.e., on a building facade or an indoor wall) and the UE is mobile (i.e., a moving vehicle, pedestrian device, or UAV). In this situation, the RIS is able to dynamically modify the phase shifts in real time to passively reflect incoming signals toward the mobile UE. The objective here is to provide sufficient signal quality to achieve a strong

and reliable communication link while the UE moves through the environment, especially when non-line-of-sight (NLOS) scenarios or obstructed areas are involved. The major challenges will be linked to quick channel estimation, real-time adjustment of the RIS settings, and maintaining continuous coverage while the UE traverses areas with continuously changing propagation conditions.

- **Mobile RIS and mobile UE**

  This refers to a dynamic communication scenario where both the RIS (e.g., drone, vehicle, robot) and the UE (mobile phone, autonomous vehicle, or drone, among others) are mobile. This scenario reflects dynamic environments (e.g., urban areas, disaster areas, or mobility networks), where such inherent flexibility enables better signal coverage through adaptive real-time beamforming. However, there are challenges with this configuration that include channel estimation, which must be fast and accurate; beam tracking, alignment, and coordination; as well as controlling energy consumption and managing complexity.

  Nevertheless, mobile RIS and mobile UE are an exciting opportunity under the paradigm of future intelligent and adaptive 6G communication systems.

2.2 RIS-aided communication system

An RIS-aided network can be a single or multi-connectivity.

2.2.1 Single RIS-aided single-user communication system

This setup can also be referred to as single-input-single-output (SISO). In the single RIS-aided single-user scenario, a single RIS is deployed to enhance the communication between a transmitter (TX), that is, BS, and a single receiver (RX), that is, UE. This configuration represents the fundamentals of the communication system [43].

Components of a single RIS-aided single-user

(1) Transmitter (BS or AP): Sends information signals to the user, either directly or via the RIS.

(2) RIS: Reflects and manipulates the incident signals to enhance wireless transmission to the user.

(3) RX: Usually a single-antenna receiver, such as a mobile phone or a sensor, that receives the signal.

Figure 1 shows the model of the single RIS-aided single user system setup network. G represents the channels from the BS to the RIS, and h represents the channels from the RIS to the user. The channels from the BS to the user are mostly blocked.

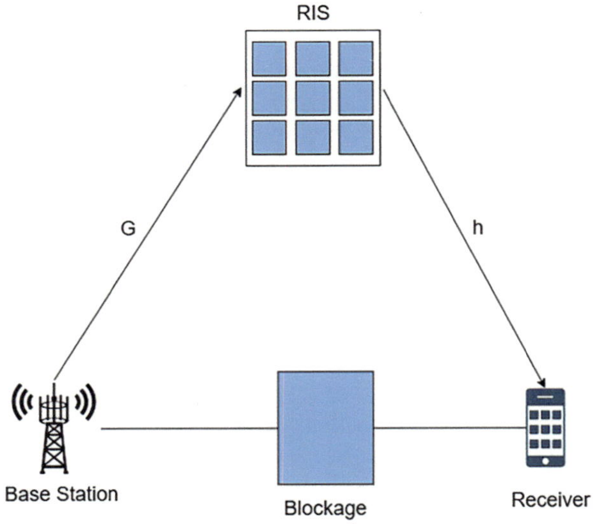

Figure 1.
Single ris-aided single-user system setup.

Channel model formulation

$$h = \mathbf{h}_{BU} + \mathbf{h}_{BR}\mathbf{h}_{RU}^{T}\Theta\mathbf{G} \tag{1}$$

where:

- h: is the effective channel from the BS (transmitter) to the user (receiver).
- h_{BR}: indicates the channel from BS to the RIS.
- h_{BU}: denotes the direct channel from the BS to the user.
- Θ: the diagonal phase shift matrix applied by the RIS. This matrix is typically represented as $\Theta = \text{diag}(e^{j\theta_1}, e^{j\theta_2}, \ldots, e^{j\theta_N})$, where θ_N is the phase shift applied by the n–th element of the RIS, and N is the number of elements in the RIS.
- $\mathbf{G}$: the channel matrix from the BS to the RIS.
- $h_{RU}^{T}\,\Theta\mathbf{G}$: represents the channel from the RIS to the user, with phase shifts applied by the RIS.

The effective channel vector from the RIS to the user incorporates the RIS-applied phase shifts. The RIS-reflected channel is a combination of three terms, namely, the BS–to-element-n channel, RIS reflection, and element-n–to-receiver channel.

$$h_{RU} = \mathbf{h}_{RU}^{T}\Theta\mathbf{G} \tag{2}$$

The signal received by the receiver can be expressed as follows:

$$y = (\mathbf{h}_{BU} + \mathbf{h}_{RU}^{T}\,\Theta\mathbf{h}_{BR})\mathbf{x} + \mathbf{n} \tag{3}$$

- x: is the transmitted signal.
- n: is the noise at the receiver.

2.2.2 Multiple RIS-aided single-user communication system

In a multiple RIS-aided single-user communication system, two or several RISs collaboratively assist a single user's communication, improving coverage, reliability, and signal strength. Using multiple RISs has significantly enhanced the performance and capacity of wireless communication by exploiting spatial diversity and multiplexing gain [31]. **Figure 2** illustrates the model for multiple RIS-aided single-user networks.

Components of a multiple RIS-aided network.

(1) TX: The source of the transmitted signal.

(2) Multiple RISs: A network with several RISs, each comprising many passive elements capable of adjusting phase shifts to reflect signals constructively toward the receiver. Each RIS is positioned to reflect and adjust the transmitted signals optimally, creating favorable signal conditions for the receiver.

(3) RX: The single receiver.

Channel model formulation

$$y = H_d + \sum_{i=1}^{M} H_{\mathrm{RIS}_i} \Phi_i H_{\mathrm{BS,RIS}_i} !x + n \tag{4}$$

where:

- x: represents the transmitted signal.

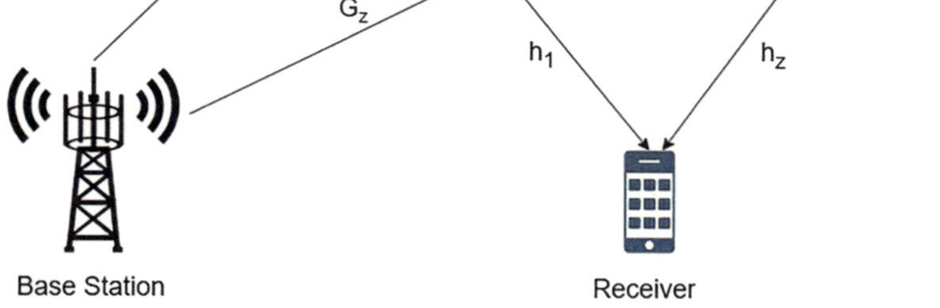

Figure 2.
Multiple ris-aided single-user system.

- n: denotes the additive noise.
- H_d: is the direct channel from the BS to the user.
- $H_{\text{RIS},i}$: is the channel matrix from RIS i to the user.
- $H_{\text{BS,RIS}i}$: is the channel matrix from the BS to RIS i.
- $\Phi_i = \text{diag}(\theta_{1i}, \theta_{2i}, ..., \theta_{Ni})$ represents the diagonal phase shift matrix of RIS_i, where θ_{ni} corresponds to the phase shift introduced by the n–th element of RIS_i.

2.2.3 Multiple RIS-aided multi-user network communication system

The multiple RIS-aided multiple-receiver system represents a sophisticated evolution in wireless communication technology, designed to enhance the quality and reliability of signal transmission. Several RISs are deployed to assist communication between a BS and multiple users. Using multiple antennas at the receiver has significantly enhanced the performance and capacity of wireless communication by exploiting spatial diversity and multiplexing gain [31]. **Figure 3** illustrates the model for a multiple RIS-aided multiple-user network.

Channel model formulation: For a system aided by multiple RISs, the received signal at the k–th user can be expressed as:

$$y_k = H_{d,k} + \sum_{i=1}^{M} H_{\text{RIS}_i} \Phi_i H_{\text{BS,RIS}_i} x + n_k, \quad k = 1, 2, ..., K \tag{5}$$

where:

- x: is the transmitted signal,
- n_k: is the noise at the k–th receiver.

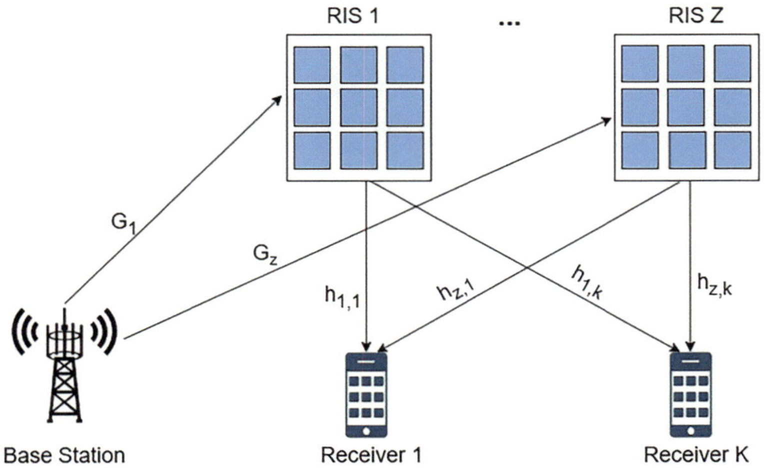

Figure 3.
Multiple ris-aided multiple-receiver system.

- $H_{d,k}$: represents the direct channel from the BS to the k–th user.
- $H_{\mathrm{BS,RIS},i}$: is the channel between the BS and RIS i.
- $H_{\mathrm{RIS},i,k}$: is the channel between RIS i and the k–th user.
- $\Phi_i = \mathrm{diag}(\theta_{1i}, \theta_{2i}, \ldots, \theta_{Ni})$: is the phase shift matrix of RIS i.
- M: is the total number of RISs.
- K: is the number of receivers.

3. Irregular RIS

While the regular RIS with uniform inter-element spacing has been extensively studied in the literature, researchers are increasingly exploring irregular RIS architectures, which offer the potential to enhance system performance by exploiting the arbitrary, non-uniform inter-element spacing as another degree of freedom (DoF) for network optimization. Irregular RIS refers to more advanced surfaces whose reflective elements are not arranged in a uniform grid or pattern, as opposed to the standard regular RIS, which consists of elements in a structured, rectangular or square grid and, therefore, may be more suitable for offering better performance in certain areas. Irregular RIS is set up with a specific number of RIS elements on an extended surface in an irregular manner, providing the RIS elements with greater flexibility. This configuration amplifies the power and system capacity to receive signals [30, 32, 33]. In practical systems, the performance of the traditional RIS is greatly influenced by the number of RIS elements [34]. Increasing the number of RIS elements leads to high costs for acquiring CSI and causes signal fading, thereby restricting the capacity of RIS-assisted communication systems [30].

Figure 4 illustrates the configuration of an irregular RIS-based network, in which a defined number of RIS elements are distributed unevenly over an extended surface.

According to the studies by Su and colleagues [32, 33], the most important features of irregular RIS arrays can be characterized as follows:

- **Non-uniform placement:** The reflective elements are distributed irregularly, either deliberately or due to design constraints, rather than being equally spaced in rows and columns.
- **Optimized configuration:** The irregular design can be optimized for specific circumstances or scenarios, which will positively improve performance measures such as signal coverage, channel capacity, and energy efficiency.
- **Design complexity:** The optimizations for the irregular array could be completed with algorithms such as Tabu Search, GA, or ML methods that could assist in finding the best arrangement for specific applications.

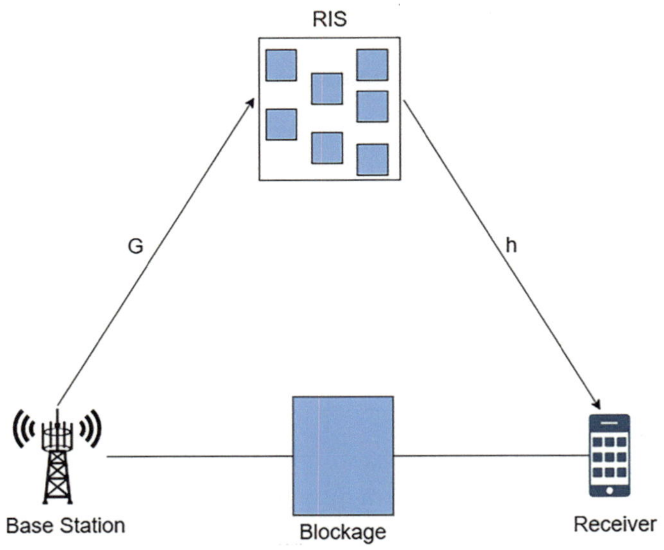

Figure 4.
Irregular RIS setup.

- **Applications:** Irregular RIS designs are particularly useful in scenarios where regular grids cannot achieve the desired results due to physical constraints or challenging environments (e.g., urban areas with obstructions).

The utilization of an irregular RIS structure in wireless communications offers several key advantages over traditional regular RIS setups:

- **Enhanced system capacity:** The irregular RIS structure allows for a more flexible arrangement of RIS elements, resulting in more degrees of freedom and greater spatial diversity. This increased flexibility leads to enhanced system capacity when using fewer RIS elements compared to a regular RIS.

- **Reduced overhead and complexity:** Unlike traditional regular RIS setups, which often necessitate extensive channel estimation and feedback due to a large number of elements, the irregular RIS structure mitigates this issue by strategically placing a smaller number of elements. This reduction in the number of elements simplifies beam design and channel management, thereby reducing system complexity.

- **Improved performance with sparse configurations:** The irregular RIS demonstrates superior performance by optimizing the sparse ratio of its elements. This optimization allows the system to maintain high performance without requiring an increased number of antennas at the BS, consequently lowering hardware costs and power consumption.

- **Flexibility in element placement:** The irregular configuration enables the selection of optimal locations for RIS elements based on channel conditions, resulting in improved signal enhancement and interference suppression. This level of flexibility is not typically available in regular RIS configurations, where elements are uniformly spaced.

4. Performance metrics

4.1 Performance metrics

Performance metrics are quantitative measures used to evaluate and analyze the effectiveness of a system. In the context of wireless communication networks, performance metrics help assess network quality, system efficiency, and user experience.

4.1.1 Signal-to-noise ratio (SNR)

It is a performance metric for applications in telecommunications, signal processing, and other fields to quantify the quality of a signal in relation to the background noise [27]. It is defined quantitatively as the ratio of the signal power (*Psignal*) to the noise power (*Pnoise*), with its unit of measure being decibels (dB). In wireless communication, a high SNR indicates that the signal power is significantly greater than the noise power, and consequently, the transmissive quality is likely to be more effective or reliable, whereas a low SNR indicates that either the noise levels are comparable to the signal levels or the noise.

Levels exceed the signal levels. In most instances, the effect of SNR is important since the quality of the transmission is hindered, and this may result in errors, degradation, and/or performance failure.

The SNR can be expressed as:

$$\mathrm{SNR} = \frac{P_{\text{signal}}}{P_{\text{noise}}}$$

In decibels (dB), it is given by:

$$\mathrm{SNR(dB)} = 10 \cdot \log_{10} \frac{P_{\text{signal}}}{P_{\text{noise}}}$$

The SNR in an RIS-assisted system is given by:

$$\mathrm{SNR} = \frac{P \cdot |h_{\text{RIS}} + h_{\text{direct}}|^2}{N_0}$$

where:

- SNR: signal-to-noise ratio (linear scale)
- P: transmit power (W)
- h_{RIS}: effective channel gain of the RIS-reflected link
- h_{direct}: channel gain of the direct link (if available)
- N_0: noise power (W)

4.1.2 Spectral efficiency (SE)

Refers to the amount of information that can be transmitted over a given bandwidth in a communication system. It measures how efficiently a system uses the available spectrum and is expressed in units of bits per second per hertz (bps/Hz) [35]. Higher SE indicates better utilization of the spectrum, enabling more data to be transmitted over a limited bandwidth. SE is a critical metric for wireless communication standards, as it impacts network capacity and user experience. RIS plays a significant role in boosting SE by optimizing the propagation environment through passive beamforming.

The general formula for SE is:

$$\mathrm{SE} = \frac{\text{Channel capacity}}{\text{bandwidth}}$$

The SE can also be calculated using the Shannon–Hartley theorem:

$$\mathrm{SE} = log_2(1 + \mathrm{SNR})$$

4.1.3 Energy efficiency (EE)

EE in communication systems measures the amount of data transmitted per unit of energy consumed [27]. EE in RIS networks refers to the ability of an RIS-assisted communication system to maximize the data rate (throughput) while minimizing power consumption. Since RIS is a passive technology, it has the potential to significantly improve the energy efficiency of wireless networks by enhancing signal quality without significantly increasing energy consumption. Higher EE indicates that a system is transmitting more data for less energy, which is desirable.

EE can be expressed as:

$$\mathrm{EE} = \frac{\text{Throughput}}{\text{Total Energy Consumption}}$$

4.1.4 Weighted sum rate (WSR)

WSR is a crucial metric utilized in wireless communication systems to assess the overall performance of a network, particularly in scenarios where users or channels are assigned different priorities or weights. This metric aggregates the individual data rates of all users within the network, adjusting them according to their respective weights to produce a single performance measure. The WSR effectively illustrates the balance between fairness and efficiency in resource allocation, highlighting the trade-offs involved in providing equitable service while maximizing overall throughput.

WSR is defined as:

$$\mathrm{WSR} = \sum_{i=1}^{N} w_i R_i$$

where:

- WSR: weighted sum rate

- N: total number of users or channels.
- w_i: weight assigned to user i ($w_i \geq 0$)
- R_i: data rate of user i.

4.2 Design considerations

Several factors influence the performance metrics of wireless communication systems. These factors can impact the overall system performance and are critical to the design and optimization of wireless networks. The key factors are:

- Number of RIS elements: The number of RIS elements plays an important role in RIS performance [36]. As the number of RIS elements increases, the beamforming gain increases, which enables even more beamforming when sending signals toward intended receivers, ultimately reducing interference and improving signal strength. The greater number of RIS elements also eases the requirements on channel estimation, enabling the RIS to provide or achieve higher throughput, reliability, and energy efficiency [37], as well as enhancing interference mitigation. However, it is worth noting that the cost associated with channel estimation and feedback does significantly increase, leading to issues with excessive computational complexity.

- Power control and transmission power: Power management and transmission levels are critical aspects that can have a direct impact on RIS performance. It's always important to consider the transmission power at the BS to ensure overall effective communication system power management [38]. When implementing efficient power management strategies, the energy needs to be appropriately allocated to each RIS element to reduce interference and enable higher received signal quality. While increased transmission power improves signal strength and SNR, excessive transmission power introduces more interference and wastes energy. Power management and transmission levels affect RIS performance, and both must be managed properly to promote signal quality, coverage, and energy efficiency.

- Element spacing and size: The physical dimensions and spacing of the meta-atoms influence the electromagnetic properties of the RIS and its ability to efficiently reflect signals. Therefore, proper design is critical to achieve maximum reflection efficiency as well as target phase shifts [39].

- Phase shift resolution: The level of resolution regarding phase shift, whether continuous or discrete, is critical to the degree of accuracy an RIS structure can achieve in signaling beams. Although higher levels of resolution typically improve overall system performance, they also add complexity in terms of hardware and costs. Hence, RIS control algorithms need to be optimized to produce the desired interference patterns while minimizing energy [40].

- Operating frequency: The selected operating frequency will affect the RIS design and operating performance; the RIS elements must be designed and configured to a specific size and shape relative to the frequency band selected, while also accounting for the required range in phase shift and response time performance dictated by that band [41].

- Hardware impairments: In reality, RIS suffers from a significant amount of hardware impairments, such as phase and amplitude errors, and limited power consumption. All of these components affect overall performance and should be taken into consideration in the design and optimization of the RIS structure. The design hardware should be scalable, low-cost, and capable of maintaining a consistent level of stability (minimizing deviation in performance between units) among the individual RIS units [42].

Figure 5 illustrates how the number of RIS elements and transmit power affect the RIS performance. It shows the sum rate (in bps/Hz) as a function of transmit power (in dBm) for different numbers of RIS elements: $N = 36$, $N = 81$, and $N = 169$. The sum rate increases as the transmit power increases, which is expected because higher power generally improves signal strength and coverage.

Also, at every transmit power level, RIS configurations with more elements (i.e., higher N) consistently achieve higher sum rates; $N = 36$ has the lowest performance: $N = 81$ performs better and $N = 169$ achieves the highest sum rate. The simulation was run for 500 iterations while sweeping the input power across a range of –70 to 20 dBm, allowing us to compute the SNR. Stochastic channel generation was used for the direct channel between the BS and the users (H_d), as well as for the channels between the BS and the RIS (G), and between the RIS and the users (H_r).

Table 1 shows the comparison of prior research on RIS.

Table A1 summarizes the principal challenges associated with RIS and presents possible solutions to address these challenges.

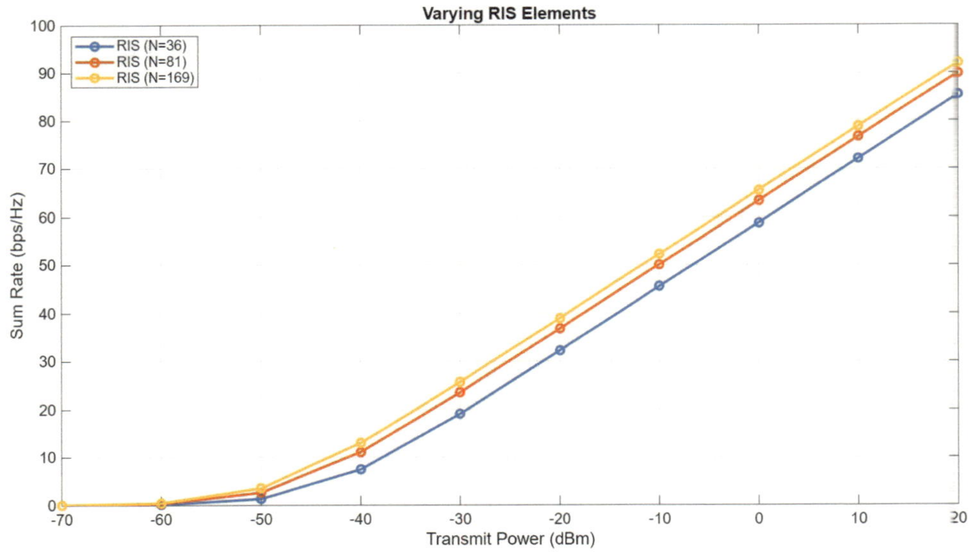

Figure 5.
Varying RIS elements.

Theme	Summary	Source	Objectives
RIS reflection, modulation, and security	Proposed a case study to demonstrate the advantages of RISs in ensuring secure communication	[43]	RIS-based techniques overview
RIS joint beamforming	A block coordinate descent (BCD) algorithm is developed to maximize the minimum rate of all users	[44]	Investigates the interplay between RIS and rate splitting by considering an RIS-assisted RS beamforming problem
RIS overview	Discusses RIS challenges, and potential solutions	[6]	An overview of the state-of-the-art in RISs, with a focus on operating principles, performance evaluation, beamforming design and resource management, machine learning applications to RIS- enhanced wireless networks, and RIS integration with other developing technologies
RIS security and privacy	Open research challenges identification and prospective future directions	[45]	RIS attributes that distinguish them from other relevant technologies

Table 1.
Comparison of prior research on RIS.

5 Conclusions

This study has provided a comprehensive understanding of RIS, covering the key optimization techniques used to configure the phase shifts. We have also presented the key performance indicators and influencing design factors that affect the RIS efficiency in practical deployments. Moreover, the deployment scenarios were highlighted in practical networks to reflect the diversity and RIS use-cases. In contrast to rigid, regular RIS structures, this study also investigated irregular designs, providing a comparative analysis of regular RIS with discussions on the benefits and drawbacks of both configurations. In summary, a holistic view of RIS was addressed, elaborating on how it can play a pivotal role in future wireless systems, including design considerations to attain the desired trade-off between optimizing throughput, coverage, energy efficiency, and complexity.

Acknowledgments

This work is supported by the European Regional Development Fund (FEDER), through the Innovation and Digital Transition Programme (COMPETE 2030) and Lisbon Regional Programme (LISBOA 2030) of the Portugal 2030 framework [Project MECON with Nr. 12950 (COMPETE2030-FEDER-00394400/LISBOA2030-FEDER-00394400)].

Nomenclature

h_{BR}	The channel from BS to the RIS
h_{BU}	The direct channel from the BS to the user
θ_n	The phase shift applied by the n-th element of the RIS

N	The number of elements in the RIS
G	The channel matrix from the BS to the RIS
h_{RU}^{T}	The channel from the RIS to the user
n	The additive noise
$H_{\mathrm{RIS},i}$	The channel matrix from RIS_i to the user
$H_{\mathrm{BS,RIS}i}$	The channel matrix from the BS to RIS i
Φ_i	The diagonal phase shift matrix of RIS i
θ_{ni}	The phase shift introduced by the nth element of RIS i
x	The transmitted signal
n_k	The noise at the k-th receiver
$H_{d,k}$	The direct channel from the BS to the k-th user
$H_{\mathrm{RIS},i,k}$	The channel between RIS i and the k-th user
M	The total number of RISs
K	The number of receivers

Abbreviations

RIS	Reconfigurable intelligent surface
UAV	Unmanned aerial vehicles
THz	Terahertz
AI	artificial intelligence
ETSI	European Telecommunications Standards Institute
ISG	Industry specification group
Los	Line-of-sight
SNR	Signal-to-noise ratio
BS	Base station
AP	Access point
AO	Alternating optimization
ML	Machine learning
DL	Deep learning
GA	Genetic algorithms
PSO	Particle swarm optimization
CSI	Channel state information
UE	User equipment
NLOS	Non-line-of-sight
SISO	Single-input-single-output
TX	Transmitter
RX	Receiver
DoF	Degree of freedom
dB	Decibels
SE	Spectral efficiency
4G	Fourth generation
5G	Fifth generation
EE	Energy efficiency
WSR	Weighted sum rate

Appendix A

Technology	Challenge category	Description	Reference	Possible solution
RIS	Channel estimation	Estimating the BS- RIS-user channel is difficult due to the RIS's nature and high training overhead.	[29, 43, 44, 46–49]	Use of active RIS elements for pilot transmission. Apply compressive sensing to reduce pilot overhead. Employ AI/ML methods
RIS	Joint beamforming	Requires coordinated optimization of BS and RIS beamforming.	[44]	Apply metaheuristic algorithms. Use AI- based optimization. Design low-complexity, scalable algorithms
RIS	Scalability	Larger RIS arrays increase hardware and algorithmic complexity.	[50]	Develop modular RIS architectures. Use low-complexity control algorithms. Invest in efficient circuit and processing architectures
RIS	Real-time control	Dynamic reconfiguration for real-time adaptation is challenging.	[6]	Implement hardware- in-the-loop optimization Use real-time learning frameworks Employ predictive control strategies
RIS	Security and privacy	Vulnerable to jamming/eaves dropping due to reconfigurability.	[45]	Integrate secure authentication mechanisms Use blockchain-based control Implement secure-by-design RIS protocols
RIS	Manufacturing and reliability	Ensuring consistent performance from large numbers of elements under environmental stress is difficult; material degradation and control electronics reliability are concerns.	[51]	Use robust materials and coatings. Develop fault-tolerant electronics. Design for redundancy and self-calibration

Table A1.
Key challenges in RIS.

Author details

Karamot Kehinde Biliaminu[1,2*†], Jonathan Rodriguez[1,3†], Felipe Gil-Castiñeira[2†] and Joaquim Bastos[1†]

1 Instituto de Telecomunicações, Universidade de Aveiro, Aveiro, Portugal

2 Universidade de Vigo, Pontevedra, Spain

3 University of South Wales, Pontypridd, United Kingdom

† These authors contributed equally

*Address all correspondence to: kehindebilly@av.it.pt

References

[1] Faisal KM, Choi W. Machine learning approaches for reconfigurable intelligent surfaces: A survey. IEEE Access. 2022;**10**:27343–27367. DOI: 10.1109/ACCESS.2022.3157651

[2] Alwarafy A, Abdallah M, Al-Dhahir N, Khattab T, Hamdi M Multi-task drl for rate control in RIS-assisted multi-cell dual-connectivity hetnets. in *Proceedings of the 2022 IEEE Global Communications Conference (GLOBECOM)*, 2022. DOI: 10.1109/globecom48099.2022.10001735.

[3] Bjornson E, Ozdogan O, Larsson EG. Reconfigurable intelligent surfaces: Three myths and two critical questions. IEEE Communications Magazine. 2020;**58**(12):90–96. DOI:10.1109/mcom.001.2000407

[4] Pan C, Ren H, Wang K, Kolb J, Elkashlan M, Chen M, Di Renzo M, Hao Y, Wang J, Swindlehurst A, You X, Hanzo L. Reconfigurable intelligent surfaces for 6G systems: principles, applications, and research directions. IEEE Communications Magazine. 2021;**59**:14–20/6. DOI: 10.1109/MCOM.001.2001076

[5] X. M, Chen Z, Chen W, Chi Y, Z. L, Han C, Wen Q. Intelligent reflecting surface enhanced indoor terahertz communication systems. Nano Communication Networks. 2020;**24**:100284. DOI: 10.1016/j.nancom.2020.100284

[6] Liu Y, Liu X, Mu X, Hou T, Xu J, Di Renzo M, Al-Dhahir N. Reconfigurable intelligent surfaces: principles and opportunities. IEEE Communications Surveys & Tutorials. 2021;**23**(3):1546–1577. DOI:10.1109/COMST.2021.3077737

[7] Georlette V, Erricolo D "Reconfigurable intelligent surfaces for industrial applications: challenges and opportunities," IECON 2024 - 50th Annual Conference of the IEEE Industrial Electronics Society, Chicago, IL, USA, 2024, pp. 1–6, DOI: 10.1109/IECON55916.2024.10905457

[8] Z. CM, Shahjalal M, Ahmed S, M JY. 6G wireless communication systems: applications, requirements, technologies, challenges, and research directions. IEEE Open Journal of the Communications Society. 2020;**1**:957–975. DOI: 10.1109/ojcoms.2020.3010270

[9] Letaief KB, Chen W, Shi Y, Zhang J, Zhang Y-JA. The roadmap to 6G: AI empowered wireless networks. IEEE Communications Magazine. 2019;**57**(8):84–90. DOI:10.1109/MCOM.2019.1900271

[10] Liu R, Zheng S, Wu Q, Jiang Y, Zhang N, Liu Y, Renzo MD, Alexandropoulos GC. Sustainable wireless networks via reconfigurable intelligent surfaces (RISs): Overview of the ETSI ISG RIS. In: IEEE Communications Standards Magazine, 2025. DOI: 10.1109/MCOMSTD.2025.3572259

[11] Eichler T, Ziegler R, *White paper: Fundamentals of THz technology for 6G*, Rohde Schwarz, 2022. [Online]. Available from: https://www.rohde-schwarz.com/solutions/test-and-measurement/wireless-communication/cellular-standards/6g/white-paper-fundamentals-of-thz-technology-for-6g-by-rohde-schwarz-registration255934.html [*Accessed*: 2025-*May*-22]

[12] Zhu Q, Gao Y, Nie J, Xiao Y, Tang W. Dynamic wireless networks assisted by RIS mounted on aerial platform: Joint active and passive beamforming design. IET Communications. 2022;**16**:1509–1522. DOI: 10.1049/cmu2.12383

[13] Feng B, Gao J, Wu Y, Zhang W, X-g. X, Xiao C. Optimization techniques in reconfigurable intelligent surface aided networks. IEEE Wireless Communications. 2021;**28**(6):87–93. DOI:10.1109/MWC.001.2100196

[14] Dong L, Wang H-M, Bai J. Active reconfigurable intelligent surface aided secure transmission. IEEE Transactions on Vehicular Technology. 2022;**71**(2):2181–2186,. DOI:10.1109/TVT.2021.3135498

[15] Zhou H, Hu C, Liu X. "An overview of machine learning-enabled optimization for reconfigurable intelligent surfaces-aided 6G networks: From reinforcement learning to large language models," arXiv, 2024. Available from: https://arxiv.org/abs/2405.17439 [Accessed: 2025-May-25]

[16] Jassim MF, Mohammed AT, Abdullah O.Deep learning-based beamforming optimization for reconfigurable intelligent surface-assisted wireless communication systems. Iraqi Journal for Computer Science and Mathematics. 2025;**6**(1). DOI: 10.52866/2788-7421.1233

[17] Huang S, Wang S, Wang R, Wen M, Huang K "Reconfigurable intelligent surface assisted edge machine learning," in *ICC 2021 - IEEE International Conference on Communications*, 2021, pp. 1–6, DOI: 10.1109/ICC42927.2021.9500445

[18] Ma H, Zhang H, Zhang W, Leung VCM. Beamforming optimization for reconfigurable intelligent surface with power splitting aided broadcasting networks. IEEE Transactions on Vehicular Technology 2023;**72**(2):2712–2717. DOI:10.1109/TVT.2022.3214529

[19] Zaid M, Kadir MKA, Shayea I, Mansor Z. Machine learning-based approaches for handover decision of cellular-connected drones in future networks: A comprehensive review. Engineering Science and Technology, an International Journal. 2024;**55**:101732. DOI: 10.1016/j.jestch.2024.101732

[20] Huang C, Mo RM, Yuen C. Reconfigurable intelligent surface assisted multiuser MISO systems exploiting deep reinforcement learning. IEEE Journal on Selected Areas in Communications. 2020;**38**(8):1839–1850. DOI:10.1109/JSAC.2020.3000835

[21] Hamadou ANS, Maina C, Soidridine MM. A hybrid PSO-GWO-based phase shift design for a hybrid-RIS-aided heterogeneous network system. Heliyon. 2024;**10**(12): e33175. DOI:10.1016/j.heliyon.2024.e33175

[22] Kennedy J, Eberhart R.Particle swarm optimization. Proceedings of ICNN'95-international Conference on Neural Networks. 1995;**27**(4):1942–1948

[23] Khaled A, Alwakeel AS, Shaheen AM, Fouda MM, Ismail MI. Placement optimization and power management in a multiuser wireless communication system with reconfigurable intelligent surfaces. IEEE Open Journal of the Communications Society. 2024;**5**:4186–4206. DOI: 10.1109/OJCOMS.2024.3426495

[24] Shami TM, El-Saleh AA, Alswaitti M, Al-Tashi Q, Summakieh MA, Mirjalili S. Particle swarm optimization: A comprehensive survey. IEEE Access. 2022;**10**:10031–10061

[25] van Zyl J-P, Engelbrecht AP.Set-based particle swarm optimisation: A review. Mathematics. 2023;**11**(13):2980

[26] Rahmat-Samii Y, Gies D, Robinson J. Particle swarm optimization (PSO): A novel paradigm for antenna designs. URSI Radio Science Bulletin. 2003;**2003**(306):14–22. DOI:10.23919/URSIRSB.2003.7909769

[27] Panuganti N, Ranjan P, Shukla A. Impact of metaheuristic optimization algorithms on wireless network coverage enhancement with reconfigurable intelligent surfaces. International Journal of Communication Systems. 2025;**38**:e70026

[28] Karthika K, Karunakaran A, Kaliappan K, Geetha D. Optimization algorithms for reconfigurable antenna design: A review. Advances in Microwave Engineering. 2023;**21**: 85-103. DOI: 10.1201/9781003459880-6

[29] Puspitasari AA, Lee BM. A survey on reinforcement learning for reconfigurable intelligent surfaces in wireless communications. Sensors. 2023;**23**(5):2554. DOI:10.3390/s23052554

[30] Zhang W, Wang L, Mao H, Wang Z, Wu W.Throughput maximization for irregular reconfigurable intelligent surface assisted NOMA systems. EURASIP Journal on Advances in Signal Processing. 2023;**2023**(1). DOI: 10.1186/s13634-023-01076-1

[31] Dajer M, Z. M, Piazzi L, Prasad N, X.-f. Q, Sheen B, Yang J, Yue G. Reconfigurable intelligent surface: Design the channel – A new opportunity for future wireless networks. Digital Communications and Networks. 2022;**8** (2):87–104. DOI:DOI: 10.1016/j.dcan.2021.11.002

[32] R. S, Dai L, Tan J, Hao M, MacKenzie R, "Capacity enhancement for irregular reconfigurable intelligent surface-aided wireless communications," in *Proc. IEEE Global Communications Conference (GLOBECOM)*, pp. 1–6, 2020. DOI: 10.1109/GLOBE-COM42002.2020.9348273.

[33] R. S, Dai L, Tan J, Hao M, MacKenzie R.Capacity enhancement for reconfigurable intelligent surface-aided wireless network: from regular array to irregular array. IEEE Transactions on Vehicular Technology. 2023;**72**(5). DOI: 10.1109/TVT.2023.3236179

[34] Yrjola S, Ahokangas P, Matinmikko-Blue M. Value creation and capture from technology innovation in the 6G Era. IEEE Access. 2022;**10**:16299–16319. DOI: 10.1109/access.2022.3149590

[35] Zhang Z. Simultaneously transmitting and reflecting reconfigurable intelligent surface-assisted non-orthogonal multiple access wireless education network under multiple interference devices. Symmetry. 2025;**17**:491. DOI: 10.3390/sym17040491

[36] Z. L, Chen W, Q. W, Wang K, Li J. Joint beamforming design and power splitting optimization in IRS-assisted SWIPT NOMA networks. IEEE Transactions on Wireless Communications. 2022;**21**(3):2019–2033. DOI:10.1109/twc.2021.3108901

[37] Wu Q, Zhang S, Zheng B, You C, Zhang R. Intelligent reflecting surface-aided wireless communications: A tutorial. IEEE Transactions on Communications. 2021;**69**(5): 3313–3351. DOI:10.1109/TCOMM.2021.3051897

[38] Galleto Jr FA, Africa ADM, Materum LY, "Impact of the number of elements and transmit power in the performance of reconfigurable intelligent surface systems," *Proceedings of the 2023 11th International Conference on Computer and Communications Management (ICCCM '23)*, pp. 144–149, ACM, New York, NY, USA, 2023. DOI: 10.1145/3617733.3617756.

[39] Alexandropoulos GC, Shlezinger N, Alamzadeh I, Imani MF, Zhang H, Eldar YC Hybrid reconfigurable intelligent metasurfaces: Enabling simultaneous tunable reflections and sensing for 6G wireless communications. arXiv 2021, arXiv:2104.04690.

[40] Pei X, Yin H, Tan L, Cao L, Li Z, Wang K, Zhang K, Bjornson E, What is SISO (Single input, single output)?. RIS-aided wireless communications: Prototyping, adaptive beamforming, and indoor/outdoor field trials. IEEE Transactions on Communications. 2021;**69**:8627–8640

[41] Garcia JCB, Sibille A, Kamoun M. Reconfigurable intelligent surfaces: Bridging the gap between scattering and reflection. IEEE Journal on Selected Areas in Communications. 2020;**38**:2538–2547

[42] Alexandropoulos GC, Shlezinger N, Del Hougne P. Reconfigurable intelligent surfaces for rich scattering wireless communications: Recent experiments, challenges, and opportunities. IEEE Communications Magazine. 2021;**59**: 28–34. DOI: 10.1002/dac.70026

[43] Xu J, Yuen C, Huang C, et al. Reconfiguring wireless environments via intelligent surfaces for 6G: Reflection, modulation, and security. Science China Information Sciences. 2023;**66**:130304. DOI: 10.1007/s11432-022-3626-5.

[44] Zhang T, Mao S. Joint beamforming design in reconfigurable intelligent surface-assisted rate splitting networks. IEEE Transactions on Wireless Communications. 2024;**23**(1):263–275. DOI:10.1109/TWC.2023.3277334

[45] Naeem F, Ali M, Kaddoum G, Huang C, Yuen C. Security and privacy for reconfigurable intelligent surface in 6g: A review of prospective applications and challenges. IEEE Open Journal of the Communications Society. 2023;**4**:1196–1217

[46] Jiang T, Cheng HV, Yu W. Learning to reflect and to beamform for intelligent reflecting surface with implicit channel estimation. IEEE Journal on Selected Areas in Communications. 2021;**39**(7):1931–1945. DOI:10.1109/JSAC.2021.3078502

[47] Chen J, Liang Y-C, Cheng HV, Yu W. Channel estimation for reconfigurable intelligent surface aided multi-user mmWave MIMO systems. IEEE Transactions on Wireless Communications. 2023;**22**(10):6853–6869. DOI:10.1109/TWC.2023.3246264

[48] You C, Zheng B, Zhang R, "Intelligent reflecting surface with discrete phase shifts: Channel estimation and passive beamforming," in *Proc. IEEE International Conference on*

Communications (ICC), 2020, pp. 1–6, DOI: 10.1109/ICC40277.2020.9149292

[49] Jensen TL, De Carvalho E, "An optimal channel estimation scheme for intelligent reflecting surfaces based on a minimum variance unbiased estimator," in *Proc. IEEE International Conference on Acoustics, Speech and Signal Processing (ICASSP)*, 2020, pp. 5000–5004, DOI: 10.1109/ICASSP40776.2020.9053695

[50] Alrikabi H, Sallomi A, KHazaal H, Magdy A. Review of the reconfigurable intelligent surfaces in smart cities: Opportunities, challenges, and applications. IET Smart Cities. 2025;**7**. DOI: 10.1049/smc2.70003

[51] Liu R, Wu Q, Di Renzo M, Yuan Y. A path to smart radio environments: An industrial viewpoint on reconfigurable intelligent surfaces. IEEE Wireless Communications. 2022;**29**(1):202–208. DOI:10.1109/MWC.111.2100258

Chapter 5

Inverted Microstrip Line (IMSL) Phase Shifter Technique for Antennas Based on Liquid Crystal

Odai Hassan Raheem Al Soad, Wan Chen and Zaid Shakir Kadhim Al-Shammari

Abstract

This chapter is devoted to potential applications of tunable phase shifters based on nematic liquid crystal (NLC) in microwave antenna structures, liquid crystal behaviors, liquid crystal characteristics, tunable structure design, simulation results, and evaluation of tunable devices. An interesting issue is designing an efficient tunable device that has the ability and flexibility to easily integrate into a microwave antenna for achieving a phase shift. The theory of liquid crystal (LC) is investigated, including its electromagnetic fields, derivation of the generalized orientation angle equation for LC molecules, the numerical method of solutions for the differential equation of the orientation angle, and the response time for LC. The orientation angle is derived and solved first for the electric field using finite-difference schemes. The solution for the differential equation of the orientation angle is obtained, which is more accurate than one approximation constant ($K_{11} = K_{22} = K_{33}$). It took 557 seconds to complete the numerical solution using the three-point difference approach. As a result, when the DC voltage increased from 0 V to 20 V, the device's differential phase shift tuned up to 57° and achieved up to 1.7 GHz frequency tuning while maintaining a low insertion loss of 1.06 dB. Meanwhile, the figure of merit value reached 76.15°/dB. The double inverted microstrip line (IMSL) phase shifter performs better than the single-IMSL phase shifter. The tunable phase shifter can be a suitable device and is easy to incorporate into microwave antennas and wireless communication systems for achieving a phase shift.

Keywords: tunable phase shifter, nematic liquid crystal, inverted microstrip line, orientation angle for LC, differential equation, finite-difference method

1. Introduction

In today's technological world, wireless communication systems play an important role, and their requirements are becoming more complex, such as in inter-satellite communications and radar systems. Tunable devices are necessary components for obtaining reconfigurable antenna systems, where the parameters of tunable devices can control the operating frequency, amplitude, or phase in microwave transceivers.

Compactness, low cost, and low-profile electronic beamforming have been studied intensively to provide tunable microwave devices with tuning elements that can be integrated into a non-tunable device [1], for example, RF-microelectromechanical systems (RF-MEMS) [2], mechanical tuning elements [3], semiconductor solutions [4], barium strontium titanium [5], and tunable dielectrics [6]. Mechanical tuning elements were commonly used in the early years of tunable microwave technology, but they result in bulky devices that are impractical. Meanwhile, electrical tuning elements can be incorporated into smaller devices with lower power consumption. Because of their low-profile characteristics for linear material, fast tuning, and high tunability, semiconductors are easy to integrate with other circuit structures or antennas that have good applications. For high frequencies, semiconductor solutions cannot meet the needs of high-frequency applications because of the weak coupling of electromagnetic wave propagation, which leads to high losses and a high profile. On the other hand, liquid crystals (LCs) and ferroelectrics are considered tunable anisotropic dielectric materials. Recent developments in the field of tunable phase shifters have led to renewed interest in LCs, which makes them suitable for high-frequency applications.

During the last decade, LCs have been utilized in tunable microwave devices, which have studied the potential microwave materials for phase shifters [7, 8], adaptive filters [9], tunable or reconfigurable antenna applications [10, 11], and polarizers [12]. Two topologies (low insertion loss and transmission line (TL)) can be classified as tunable phase shifters depending on the technology. There are several available methods for tuning metamaterials, including geometry alteration [13], diodes [14], and electro-optic (EO) tuning [15]. Excellent tunable phase shifter materials are LC and varactor diode materials because of their low power consumption and low voltage control, which provide a large tuning range. In addition, because of their broad range of operating frequencies or large dielectric changes and effects, the linear or nonlinear properties of metamaterials are highly advantageous.

Liquid crystals, or more precisely nematic liquid crystals (NLCs), are today the dominant technology used for display applications, owing to the controllable anisotropy of the material, their low dielectric loss, efficiency, high linearity, and high and continuous tunability. The LC devices have been promised to provide tunable materials for microwave and millimeter-wave applications. Unfortunately, LC devices suffer from limited temperature stability [16]. Liquid crystals exhibit a mesophase between the solid crystalline phases and the isotropic liquid phase. The mesophase can take the nematic, smectic, or cholesteric form, depending on the orientation and positional order of the molecules of the LC. Different topologies can be used for tunable anisotropic dielectrics. Inverted microstrip line (IMSL) is a typical type of phase shifter of TL topology due to its low insertion loss (IL) for tunable anisotropic dielectrics. The IMSL uses NLC as the effective substrate for controlling the phase delay of a transverse electromagnetic mode (TEM) waveguide. The molecules of LC in the nematic phase show no positional order. The Fredericksz effect causes the molecules of LC to exhibit a long-range orientation order by using a static or low-frequency external electric field or magnetic field. The permittivity or refractive index of NLC can be changed by orienting the one-dimensional director of the molecules.

Tunable phase shifters based on LC are an important key to realizing phased arrays and reflect arrays in reconfigurable antenna systems, which are usually operated at frequency ranges above 10 GHz and up to 100 GHz. For example, flexible inter-satellite communication (geostationary satellite functions) must constantly track the trajectory of low Earth orbit (LEO) satellites, which previously could only be achieved using mechanical beam steering that required space within the antenna body. As a result, electrical tuning elements provide a solution for reducing the weight of the antenna. On the other hand, automotive radar can be realized using phased arrays, which strongly require highly configurable speed. These factors make LC a promising tunable element in the future of wireless communication systems.

In this chapter, the orientation angle of LC molecules serves as the key player in changing the permittivity of LC, which is controlled by an external electric or magnetic field. The orientation angle equation can be derived as a differential equation for the molecule director, its proper functioning, and its stability in user-defined positional order, which are necessary and important to develop an accurate and fast tunability differential equation. Thus, numerical solutions of the differential equation for the LC molecules' orientation angle certainly need to link up with the electric field. An inaccurate assessment of one constant approximation ($K_{11} = K_{22} = K_{33}$) could lead to a large deviation in the orientation angle of LC molecules between theoretical and practical results. A more critical concern is that it forces numerical solutions with the electric field. For tunable phase shifter applications, tunable antennas based on LC are getting less attention due to difficulties in storing LC during the experimental process, which could be one reason, but the emergence and utilization of designing a simple and effective phase shifter have also led to this work. Two IMSL phase shifter structures (novel double IMSL and common single IMSL) will be investigated to determine that the double-IMSL phase shifter shows a higher performance than a single-IMSL phase shifter with a differential phase shift of up to 57°.

2. Fundamentals of LC

As it is known, LC materials can be categorized into different phases or states in nature, such as solid, liquid, and gas. Basically, strong intermolecular forces can cause molecules to stay together in a solid-state material. In another way, molecules can be organized in certain positions that can orient in a certain direction. The properties of orientation and positional order can be realized in crystals. Unlike solid-state materials, liquid-state materials have weaker intermolecular forces, which make their molecules movable and able to fill the same container form but without any regular arrangement. Similarly, gas-state material has the same properties as liquid-state material, but its molecules are separated throughout the entire container. The phase of a material can be changed depending on its temperature and pressure. The mesophase of the LC can lie between the solid state and liquid state, where its molecules can flow like a liquid but have a certain arrangement like a crystal. Molecules of the LC may take a discotic shape or a calamitic shape, with the latter playing an important role in utilizing tunable microwave devices.

The phase of the LC depends on the temperature (T), which is referred to as thermotropic LC. The LC phase is determined by the melting temperature point (T_m) and the clearing temperature point (T_c). Below Tm, the LC phase will be in a solid state with both positional and orientational order. Above T_c, the LC phase will be

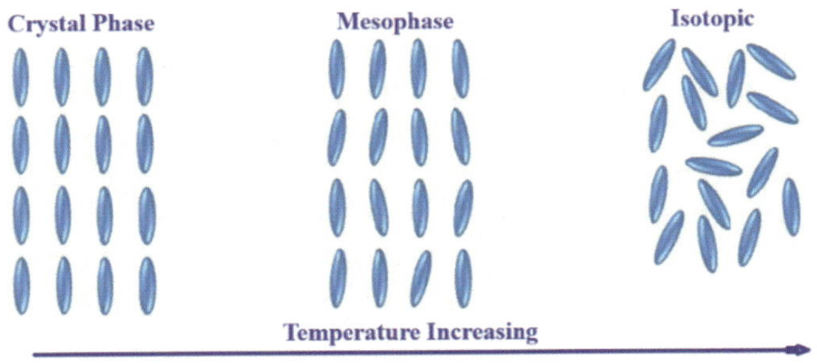

Figure 1.
The schematic behavior of LC molecules with temperature.

entirely isotropic, exhibiting no orientational or positional order. Meanwhile, between Tm and Tc, the LC phase exhibits no positional order, similar to an isotropic liquid, but demonstrates orientational order. This state of LC is called the NLC phase, as shown in **Figure 1**. When discussing LC, the director *(n)* and the order parameter (*S*) need to be explained. The director *(n)* of an LC molecule is considered a unit vector, along which all LC molecules can be described in the direction of their longer axis. The orientation of LC molecules aligns in a direction that represents the same time-averaged direction.

On the other hand, the order parameter (*S*) represents quantitatively how the LC is arranged. The value of *S* must be 1 when it represents highly ordered materials, such as a solid crystalline structure. Meanwhile, there is no orientational order when the value of the order parameter is 0, as in an isotropic liquid. The order parameter (*S*) can be expressed as [17]:

$$S(T) = P_2\, cos(\theta) = \frac{1}{2}\left(3\, cos^2(\theta)\right) = \frac{\int_0^{\pi}\left(\frac{1}{2}\left(3\, cos^2(\theta) - 1\right) f(\theta, T)\, sin(\theta)\right) d\theta}{\int_0^{\pi}\left(f(\theta, T)\, sin(\theta)\right) d\theta} \tag{1}$$

Where P_2 – the second-order Legendre polynomial,
θ – the orientation angle of LC molecules,
$f(\theta)$ — the orientation distribution function.

The LC depends only on the orientation polar angle θ, which the LC refers to as a uniaxial material [17]. Meanwhile, the value of *S* depends on the temperature, which is described in detail by the Landau-de Gennes and Maier-Saupe theory [17].

3. Electromagnetic properties of LC

3.1 Dielectric properties of LC

As dielectric materials, the relative permittivity (εr) is an important property of LC for microwave applications. The relative permittivity of LC is linked to the polarizability of the dielectric materials, which is in response to the externally applied electric field. Moreover, the applied electric polarization (*P*) depends on the internal field constant tensor ($\hat{K}$) and the LC's molecular material polarization tensor ($\hat{\alpha}_p$), which can be defined as [18]:

$$\boldsymbol{P} = N\ \hat{\alpha}_{\mathrm{p}}\hat{K}\boldsymbol{E} \tag{2}$$

where N – The molecular number density.

Meanwhile, the internal field constant tensor and molecular material polarization tensor can be expressed as [19]:

$$\hat{K} = \begin{bmatrix} K_{\perp} & 0 & 0 \\ 0 & K_{\perp} & 0 \\ 0 & 0 & K_{\parallel} \end{bmatrix} \text{ and } \hat{\alpha}_{\mathrm{p}} = \begin{bmatrix} \alpha_{\mathrm{p},\perp} & 0 & 0 \\ 0 & \alpha_{\mathrm{p},\perp} & 0 \\ 0 & 0 & \alpha_{\mathrm{p},\parallel} \end{bmatrix} \tag{3}$$

where $\perp$ – the perpendicular to the molecular axis,

$\parallel$ –_ the parallel to the molecular axis.

The Eqs. (2) and (3) are aligned with the molecular axis (a). Therefore, these tensors could be represented in xyz coordinates by using a rotation matrix (R):

$$R = \begin{bmatrix} \cos(\theta)\cos(\phi) & -\sin(\phi) & \sin(\theta)\cos(\phi) \\ \cos(\theta)\sin(\phi) & \cos(\phi) & \sin(\theta)\sin(\phi) \\ -\sin(\theta) & 0 & \cos(\theta) \end{bmatrix} \tag{4}$$

So obtained:

$$\vec{P} = N\ R\ \hat{\alpha}_p\hat{K}\ R^{-1}\vec{E} \tag{5}$$

Similarly, the electric polarization (P) can be expressed in relation to the electric field (E) in another equation as [20, 21]:

$$\boldsymbol{P} = \varepsilon_o\chi_e\boldsymbol{E} \tag{6}$$

where χ_e – the electric susceptibility for anisotropic materials,

$\hat{\chi}_e$ – the electric susceptibility tensor.

By multiplying R, R-1, $\hat{\alpha}_{\mathrm{p}}$, and $\hat{K}$ in Eq. (5) and comparing with Eq. (6). Clearly, the electric susceptibility depends on the orientation angle (θ) of each LC molecule. As a result, the electric susceptibility tensor can be rewritten after cancelling the variable ϕ from Eq. (4) as [22, 23]:

$$\widehat{\chi_{\mathrm{e}}} = \frac{N}{3\mathcal{E}_o}\begin{bmatrix} \alpha_{\mathrm{p},\perp}K_{\perp}(2+S)+\alpha_{\mathrm{p},||}K_{||}(1\text{-}S) & 0 & 0 \\ 0 & \alpha_{\mathrm{p},\perp}K_{\perp}(2+S)+\alpha_{\mathrm{p},||}K_{||}(1\text{-}S) & 0 \\ 0 & 0 & \alpha_{\mathrm{p},\perp}K_{\perp}(2\text{-}2S)+\alpha_{\mathrm{p},||}K_{||}(1+2S) \end{bmatrix} \tag{7}$$

Relative permittivity can be determined by $\varepsilon_{\mathrm{r}} = 1 + \chi_{\mathrm{e}}$ for a material by knowing the electric susceptibility, and this relation is valid for isotropic materials only in a scalar form. On the other hand, the dielectric displacement (D) can be defined for anisotropic materials as [21]:

$$\boldsymbol{D} = \varepsilon_{\mathrm{o}}\boldsymbol{E} + \boldsymbol{P} = \varepsilon_{\mathrm{o}}\left(\hat{I} + \hat{\chi}_{\mathrm{e}}\right)\boldsymbol{E} = \varepsilon_o\hat{\varepsilon}_{\mathrm{r}}\boldsymbol{E} \tag{8}$$

where $\hat{I}$– the identity matrix tensor.

Obtained:

$$\widehat{\mathcal{E}_r}=\left(\hat{I}+\widehat{\chi_e}\right)=\begin{bmatrix} 1+\frac{N}{3\mathcal{E}_o}(\alpha_{p,\perp}K_{\perp}(2+S)+\alpha_{p,\parallel}K_{\parallel}(1-S)) & 0 & 0 \\ 0 & 1+\frac{N}{3\mathcal{E}_o}(\alpha_{p,\perp}K_{\perp}(2+S)+\alpha_{p,\parallel}K_{\parallel}(1-S)) & 0 \\ 0 & 0 & 1+\frac{N}{3\mathcal{E}_o}(\alpha_{p,\perp}K_{\perp}(2-2S)+\alpha_{p,\parallel}K_{\parallel}(1+2S)) \end{bmatrix} \quad (9)$$

Based on Eq. (9), two different values of relative permittivity can be written as:

$$\varepsilon_{r\perp} = 1+\frac{N}{3\varepsilon_o}\left(\alpha_{p,\perp}K_{\perp}(2+S)+\alpha_{p,\parallel}K_{\parallel}(1-S)\right) \quad (10)$$

and

$$\varepsilon_{r\parallel} = 1+\frac{N}{3\varepsilon_o}\left(1+\alpha_{p,\perp}K_{\perp}(2-2S)+\alpha_{p,\parallel}K_{\parallel}(1+2S)\right) \quad (11)$$

The anisotropy can be obtained by taking the difference between Eq. (11) and Eq. (10), as [22]:

$$\Delta\varepsilon_r = \varepsilon_{r\parallel} - \varepsilon_{r\perp} = \frac{N}{\varepsilon_o}\left(\alpha_{p,\parallel}K_{\parallel} - \alpha_{p,\perp}K_{\perp}\right)S \quad (12)$$

Equation (12) is proportional to the value of the order parameter. The permittivity of a material may be of a complex value. The imaginary part can be represented by the material losses or the heat dissipation, which is caused by vibration damping. These losses cannot be characterized by the mechanism of conductor losses, which are determined by material conductivity (σ). It can be obtained from Maxwell's equation as:

$$\nabla \times \boldsymbol{H} = j\omega\boldsymbol{D} + \boldsymbol{J} = j\omega\left(\varepsilon' - j\left(\varepsilon'' + \frac{\sigma}{\omega}\right)\right)\boldsymbol{E} \quad (13)$$

The loss tangent ($\tan\delta$) is the ratio between the imaginary part and real part, which is used to describe the losses for the material in microwave applications as [21]:

$$\tan\delta = \frac{\omega\,\varepsilon'' + \sigma}{\omega\,\varepsilon''} \quad (14)$$

For LC, the loss tangent can also be shown to have anisotropic properties as reported in [18]. Typically, the perpendicular state of loss tangent is higher than the parallel state of it. The reason for using LC in high-frequency applications (above 10 GHz) is because the loss tangent decreases with increasing frequency [23]. Finally, the perpendicular and parallel permittivity are aligned to the molecule director, while the electric field is aligned to the xyz coordinate. Therefore, the permittivity tensor ($\hat{\varepsilon}_r$) in the xyz coordinate is:

$$\hat{\varepsilon_{\rm r}} = R\begin{bmatrix} \varepsilon_{\rm r\perp} & 0 & 0 \\ 0 & \varepsilon_{\rm r\perp} & 0 \\ 0 & 0 & \varepsilon_{\rm r\|} \end{bmatrix} R^{-1} = \begin{bmatrix} \varepsilon_{\rm rxx} & \varepsilon_{\rm rxy} & \varepsilon_{\rm rxz} \\ \varepsilon_{\rm rxy} & \varepsilon_{\rm ryy} & \varepsilon_{\rm ryz} \\ \varepsilon_{\rm rxz} & \varepsilon_{\rm ryz} & \varepsilon_{\rm rzz} \end{bmatrix} \tag{15}$$

3.2 Magnetic properties of LC

For LC material, it also offers anisotropic magnetic properties. By applying a magnetic field (H) on the LC material, the LC molecules can be aligned parallel to the magnetic field. The magnetic polarizability ($\hat{\kappa}$) for LC molecules can be expressed as [18]:

$$\hat{\kappa} = \begin{bmatrix} \kappa_\perp & 0 & 0 \\ 0 & \kappa_\perp & 0 \\ 0 & 0 & \kappa_\| \end{bmatrix} \tag{16}$$

where $\kappa_\perp$ – the perpendicular magnetic polarizability,
$\kappa_\|$ – the parallel magnetic polarizability.

Equation (15) is represented in the molecule axis, which also can be represented in the xyz axis by using the rotation matrix from Eq. (4). Similar to the electric properties, the magnetic susceptibility tensor ($\hat{\chi_{\rm m}}$) is expressed as:

$$\hat{\chi_{\rm m}} = N\hat{\kappa} = N\begin{bmatrix} \kappa_\perp + \frac{1}{3}(1-S)\Delta\kappa & 0 & 0 \\ 0 & \kappa_\perp + \frac{1}{3}(1-S)\Delta\kappa & 0 \\ 0 & 0 & \kappa_\perp + \frac{1}{3}(2S-1)\Delta\kappa \end{bmatrix} \tag{17}$$

and

$$\Delta\kappa = \kappa_\| - \kappa_\perp \tag{18}$$

As a result, the anisotropic can be obtained as [22]:

$$\Delta\chi = \chi_{\rm m\|} - \chi_{\rm m\perp} = N\Delta\kappa\, S \tag{19}$$

χ is usually a positive value, while$\chi_{\rm m\perp}$and $\chi_{\rm m\|}$ are negative and small values (their magnitude is around 10^{-5} [22]). Therefore, the relative permeability tensor ($\hat{\mu}$) can be defined as:

$$\hat{\mu_r} = \hat{\chi_m} + \hat{I} \tag{20}$$

This anisotropism is negligible. The magnetic property can be referred to as the magnetic alignment of LC molecules. As a result, the orientation of LC molecules changes so that the relative permeability is altered by the microwave field.

3.3 Benchmark values of the tunable material

The tunability ($\tau_{\rm LC}$) is characterized by the dielectric performance of a tunable material, which can be expressed for NLC as:

$$\tau_{LC} = \frac{\varepsilon_{r||} - \varepsilon_{r\perp}}{\varepsilon_{r||}} \tag{21}$$

The Eqs. 2–22 assumed that $\varepsilon_{r||} > \varepsilon_{r\perp}$, mostly LCs, have positive anisotropy, which means $\varepsilon_r > 0$ and $\tan\delta\perp > \tan\delta||$. **Figures 2** and **3** explain continuously tuning dielectric permittivity and loss tangent versus the bias voltage in a typical LC layer. The material efficiency of LC (η_{LC}) can be expressed according to the dielectric loss tangent as:

$$\eta_{LC} = \frac{\tau_{LC}}{\max\,(\tan\delta)} \tag{22}$$

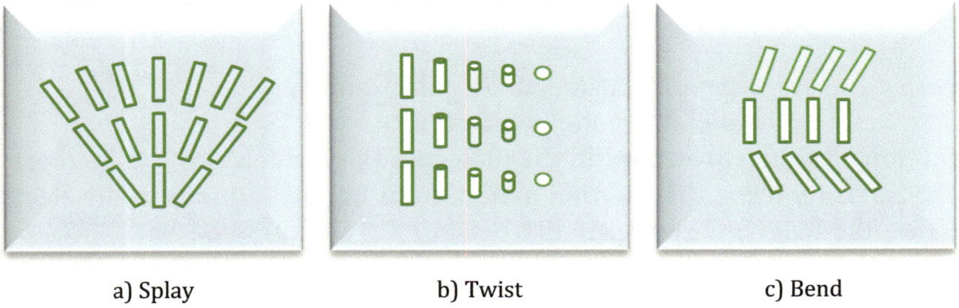

Figure 2.
Elastic deformation of the LC.

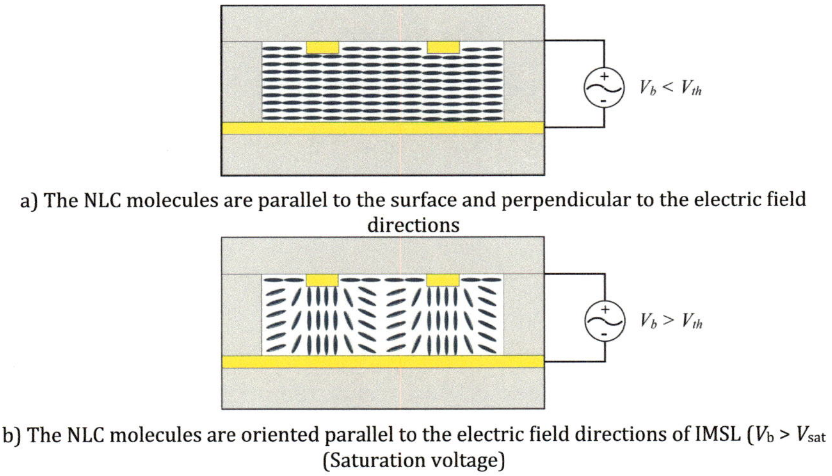

Figure 3.
The cross-section views of the IMSL. (a) The NLC molecules are parallel to the surface and perpendicular to the electric field directions; (b) The NLC molecules are oriented parallel to the electric field directions of IMSL ($v_b > v_{sat}$, saturation voltage).

The permeability of LC is approximately non-varying ($\mu_r \approx 1$) [24]. In optical frequencies, the index of refraction is used instead of permittivity, and the optical anisotropy (Δn) can be defined as:

$$\Delta n = n_{\|} - n_{\perp} = \sqrt{\varepsilon_{r\|}} - \sqrt{\varepsilon_{r\perp}} \tag{23}$$

4. Theoretical analysis of LC

4.1 The Oseen-Frank energy

The LC theory is considered a continuum of energy formulations that fills the entire cavity and is distributed continuously, possessing elastic properties based on the director n. This energy can be described by the Gibbs energy type, which should be minimized at a constant temperature and pressure. When there is a change in the orientation of the director, the energy will increase [22, 23]. In NLC, three elastic deformation types of directors are observed due to external forces. **Figure 2** illustrates that the basic deformations are splay, twist, and bend.

According to the nature of these elastic deformations, the total system's energy contribution for each deformation type can be rewritten as:

$$f_{elestic} = f_{elestic} + f_{twist} + f_{bend} = \frac{1}{2}K_{11}(\nabla.n)^2 + \frac{1}{2}K_{22}\left(n.(\nabla.n)^2\right) + \frac{1}{2}K_{33}\left(n.(\nabla.n)^2\right) \tag{24}$$

where K_{11} – the splay deformation constant,
K_{22} – the twist deformation constant,
K_{33} – the bend deformation constant (the Oseen-Frank elastic constant).

4.2 Response to the electric and magnetic fields

In the following, an electric field can be used to orient the bulk of the LC. In order to obtain the theory of response to an external electric field, the energy density can be calculated for the electric field as [25]:

$$f_{electric} = \frac{1}{8\pi}\boldsymbol{E}.\boldsymbol{D} = \frac{\varepsilon_r}{8\pi}\boldsymbol{E}^2 + \frac{\Delta\varepsilon_r n.\boldsymbol{E}^2}{8\pi} \tag{25}$$

The free energy of LC is influenced by another factor called anchoring energy, which arises due to contact with the LC's container. Based on the Rapini-Papoular model, the free energy of LC can be expressed as [22]:

$$f_{surface} = \frac{1}{2}W_p \sin^2(\theta - \theta_0) + \frac{1}{2}W_a \sin^2(\phi - \phi_0) \tag{26}$$

where W_P – the polar anchoring strength,
W_a – the azimuthal anchoring strength,
Θ_0 & ϕ_0 — the preferred orientation.

In general, the polar anchoring is larger than the azimuthal anchoring by one or two orders of magnitude [26]. The electric and magnetic fields can make a higher contribution to the Gibbs free energy. Therefore, the free energy can be written per unit of volume as [23]:

$$f_{\text{electric}} = \frac{1}{2}\varepsilon_0 \left(\boldsymbol{E}^2\right)\left(\varepsilon_{\mathrm{r}\perp} + \Delta\varepsilon_{\mathrm{r}} \cos^2\vartheta\right) \tag{27}$$

Similarly, the magnetic field contribution can be obtained as:

$$f_{magnetic} = \frac{1}{2}\mu_0 \left(\boldsymbol{H}^2\right)\left(\mu_{r\perp} + \Delta\mu_r\ cos^2\vartheta\right) \tag{28}$$

where ϑ – the angle between the director and the electric field.

Therefore, the total Gibbs free energy can be expressed with all these contributing factors as [22]:

$$\mathcal{F} = \int_{vol} \left(f_{\text{elestic}} + f_{\text{surface}} - f_{\text{electric}} - f_{\text{magnetic}}\right) dv \tag{29}$$

The behavior of the LC molecule director is governed by the effective value of the external electric field. The system can be realized by a specific LC and an electric field that can adjust the director's orientation. Equation (29) is minimized as:

$$\mathcal{F} = \int_{vol} \left(f_{\text{elestic}} - f_{\text{electric}}\right) dv \tag{30}$$

The external electric field (f_{electric}) is considered negative to align with the definition of Gibbs free energy. According to this understanding, the system demonstrates the tuning of NLC directions to a parallel orientation when the bias voltage is applied. The applied voltage is connected between parallel plate capacitors, which are filled with NLC. However, if the director is accidentally distorted, the elastic energy will increase, which also raises the total Gibbs free energy.

The phase shifter structure consists of the IMSL and a ground plane, which allow the electric field directions to be distributed inside the LC cell. A certain bias voltage is applied to the IMSL. Basically, two voltages, fixed (V_0) and tunable (V_b), can reach the perpendicular state and parallel state of LC orientation by tuning the bias voltage from $0V$ to V_b. The electric biasing system of the structure is illustrated in **Figure 3**. The IMSL and ground plane are considered as a large parallel plate capacitor. The equipotential lines will be parallel to the IMSL, which makes the electric field perpendicular to the IMSL and the ground plane. This concept of IMSL tuning can reduce the performance of the structure, which should be obvious in high-frequency fields. In general, when the configuration of the perpendicular state is set, the current will flow in the IMSL and ground plane due to the voltage difference of 0 V and V_b. Obviously, it will take an infinite biasing voltage to fully reorient the LC molecules' directors.

Nevertheless, it is a more important factor to know how much biasing voltage is needed to achieve 90% or 95% tunable LC. Moreover, the dissipated power can change the property of the LC inside the structure due to self-heating, and unwanted power dissipation occurs

4.3 Derivation of generalized orientation angle equation

The NLC has three fundamental types of deformation, and those types are called splay, twist, and bend deformations. Each one of these deformation types can contribute to the system's energy as the Oseen-Frank energy, and the elastic energy is combined with these three deformation types (Eq. (24)). The total system's energy is defined by a specific elastic energy of LC and an electric field, which can be adjusted to the orientation director n as:

$$\mathcal{F} = \int \begin{pmatrix} \frac{1}{2}K_{11}(\nabla \cdot \boldsymbol{n})^2 + \frac{1}{2}K_{22}\left(\boldsymbol{n} \cdot (\nabla \times \boldsymbol{n})^2\right) + \frac{1}{2}K_{33}\left(\boldsymbol{n} \times (\nabla \times \boldsymbol{n})^2\right) \\ -\frac{1}{2}\varepsilon_0 \boldsymbol{E}^2 \varepsilon_{r\perp} + \Delta\varepsilon_r(\boldsymbol{n}.\boldsymbol{E})^2 \end{pmatrix} d^3r \tag{31}$$

By using the Euler-Lagrange equation, the functional ($\mathcal{F}$) in Eq. (31) can be minimized as:

$$\frac{\partial f}{\partial \theta} - \frac{d}{dz} \cdot \frac{\partial f}{\partial \theta_z} \tag{32}$$

and

$$f = \frac{1}{2}K_{11}(\nabla .\boldsymbol{n}\ )^2 + \frac{1}{2}K_{22}\left(\boldsymbol{n} \cdot (\nabla \times \boldsymbol{n}\ )^2\right) + \frac{1}{2}K_{33}\left(\boldsymbol{n} \times (\nabla \times \boldsymbol{n}\ )^2\right) - \frac{1}{2}\varepsilon_0 \boldsymbol{E}^2 \varepsilon_{r\perp} + \Delta\varepsilon_r(\boldsymbol{n} \cdot \boldsymbol{E})^2 \tag{33}$$

where $\theta_z = \frac{\partial \theta}{\partial z}$
By manipulating the individual terms in Eq. (33) as follows:

$$f = \frac{1}{2}\ K_{11}\ \theta_z^2\ \sin^2\theta + \frac{1}{2}\ K_{33}\ \theta_z^2\ \cos^2\theta - \frac{1}{2}\ \varepsilon_0\ \boldsymbol{E}^2\ \varepsilon_{r\perp} + \Delta\varepsilon_r\ \cos^2\vartheta \tag{34}$$

Differentiating the resulting terms of Eq. (34) as:

$$\frac{\partial f}{\partial \theta} = \frac{1}{2}(K_{11} - K_{33})\sin 2\theta\left(\frac{\partial \theta}{\partial z}\right)^2 + \frac{1}{2}\varepsilon_0\ \Delta\varepsilon_r \boldsymbol{E}^2 \sin 2\theta \tag{35}$$

$$\frac{d}{dz} \cdot \frac{\partial f}{\partial \theta_z} = (K_{11} - K_{33}\ )\ \sin 2\theta\left(\frac{\partial \theta}{\partial z}\right)^2 + \left(K_{11}\ \sin^2\theta - K_{33}\ \ \cos^2\theta\right)\frac{\partial^2 \theta}{\partial z^2} \tag{36}$$

By substituting Eqs. (35) and (36) into Eq. (32), the following is yielded:

$$\frac{1}{2}\ \varepsilon_0\ \Delta\varepsilon_r\ \boldsymbol{E}^2\ \sin 2\theta - \frac{1}{2}\ (\ K_{11} - K_{33}\ )\ \sin 2\theta\left(\frac{\partial \theta}{\partial z}\right)^2 - (\ K_{11}\ \sin^2\theta - K_{33}\ \ \cos^2\theta)\ \frac{\partial^2 \theta}{\partial z^2} = 0 \tag{37}$$

Rearranging Eq. (37) yields:

$$\frac{1}{2}\ \frac{\varepsilon_0\ \Delta\varepsilon_r\ \boldsymbol{E}^2\ \sin 2\theta}{(K_{11}\ \sin^2\theta + K_{33}\ \ \cos^2\theta)} - \frac{1}{2}\ \frac{(K_{11} - K_{33}\ )\ \sin 2\theta}{(K_{11}\ \sin^2\theta + K_{33}\ \ \cos^2\theta)}\ \left(\frac{\partial \theta}{\partial z}\right)^2 - \frac{\partial^2 \theta}{\partial z^2} = 0 \tag{38}$$

Based on Eq. (38), the Freedericksz transition voltage (V_{th}) is a voltage at which no molecular reorientation occurs and can be reduced, which can be defined as:

$$V_{th} = \frac{\pi}{h_c}\sqrt{\frac{K_{11}}{\varepsilon_0\ \Delta\varepsilon_r}} \tag{39}$$

where h_c __ The thickness of the LC.

Equation (37) is denoted only for the configuration in **Figure 3**, which is dominated by splay deformation only (K_{11}). Meanwhile, if it is dominated by twist deformation, this equation will involve any type of director distortion, which is more complex. Equation (38) can be rewritten as:

$$\frac{1}{2}\left(\frac{V}{V_{th}}\right)^2\left(\frac{\pi}{h_c}\right)^2\frac{\sin 2\theta}{(K_{11}\ \sin^2\theta + K_{33}\ \cos^2\theta)} - \frac{1}{2}\frac{(K_{11} - K_{33}\ )\ \sin 2\theta}{(K_{11}\ \sin^2\theta + K_{33}\ \cos^2\theta)}\left(\frac{\partial\theta}{\partial z}\right)^2 - \frac{\partial^2\theta}{\partial z^2} = 0 \tag{40}$$

The differential equation in Eq. (40) is more complex and even difficult to handle numerically. Therefore, in [27], it was assumed that the elastic deformation constants are equal ($K_{11} = K_{22} = K_{33}$), which is called the one-constant approximation. Based on this proposal, Eq. (40) can be minimized as shown in [28, 29]:

$$\frac{1}{2}\left(\frac{V}{V_{th}}\right)^2\left(\frac{\pi}{h_c}\right)^2 \sin 2\theta - \frac{\partial^2\theta}{\partial z^2} = 0 \tag{41}$$

The differential Eq. (40) is derived from a consideration of an accurate equation for utilizing the orientation angle of the NLC molecule rather than Eq. (41). Therefore, Eq. (40) can be solved iteratively using the finite-difference method of solution.

The accurate method of solution is to subdivide the thickness of the LC cell into N-layers, which can be assumed to make the LC a homogeneous anisotropic medium, as shown in **Figure 4**. The thickness of each layer is assumed to be h_L as a relaxation factor. The external electric field exists in three separate regions: region 1 is the top plane ($z \geq h_c$), region 2 is the LC cell ($0 \leq z \leq h_c$), and region 3 is the ground plane ($z \leq 0$).

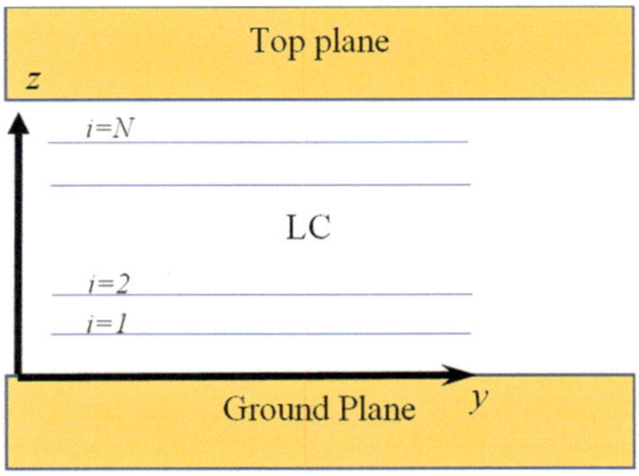

Figure 4.
The subdivided solution geometry.

This method of solution is obtained analytically for the external electric field. The director angle of the NLC molecule can be determined at any point inside the LC cavity. The nonlinear equation of the director angle is solved iteratively using a finite-difference scheme for the electric field.

The process of iteration is repeated until convergence is achieved. A two-point boundary can be defined by the orientation directors versus the z-coordinate, and it is controlled by Eq. (40). Dirichlet type is used for the two-point boundary conditions of strong anchoring as $\theta(z=0, z=hc)=0$. Robin type can be expressed in the boundary conditions as a first-order differential equation as:

$$\left(K_{11}\,\sin^2\theta - K_{33}\,\cos^2\theta\right)\left(\frac{d\theta}{dz}\right) + \frac{1}{2}\left(\frac{dF}{d\theta}\right) = 0 \quad at\ z = 0, h_c \tag{42}$$

where $F(\theta) = \ C\cos^2\theta + C_4\cos^4\theta$ __ the interfacial potential.

By letting $h_L= d/2$, the differential equation in Eq. (40) is discretized as:

$$\frac{1}{2}\left(\frac{V}{V_{th}}\right)^2\left(\frac{\pi}{h_c}\right)^2\frac{\sin 2\theta}{\left(K_{11}\,\sin^2\theta + K_{33}\,\cos^2\theta\right)} - \frac{1}{8}\,\frac{\left(K_{11} - K_{33}\right)\sin 2\theta}{\left(K_{11}\,\sin^2\theta + K_{33}\,\cos^2\theta\right)}\left(\theta_{i+1} - \theta_{i-1}\right)^2 - \left[\theta_{i+1} - 2\theta_i + \theta_{i-1}\right] = 0 \tag{43}$$

For greater stability in the algorithm of the three-point finite-difference scheme with relaxation, the convergence rate is:

$$\theta_i^{k+1} = \frac{1}{3}\left(\frac{1}{2}\left(\frac{V^2}{V_{th}^2}\right)\left(\frac{\pi}{h_c}\right)^2\frac{h_L^2\,sin2\theta_i^k}{\left(K_{11}\,sin^2\theta_i^k + K_{33}\,cos^2\theta_i^k\right)} - \frac{\left(K_{11} - K_{33}\right)sin2\theta_i^k}{\left(K_{11}\,sin^2\theta_i^k + K_{33}\,cos^2 q_i^k\right)}\left(\theta_{i+1}^k - \theta_{i-1}^k\right)^2 - \left(\theta_{i+1}^k - 2\theta_i^k + \theta_{i-1}^k\right)\right) \tag{44}$$

In order to provide soft anchoring, the boundary conditions of the Robin type are imposed. Therefore, Eq. (42) is discretized as:

$$\theta_i^{k+1} = \theta_{i+1}^k - h_L\left[\frac{\sin\theta_i^k\left(C\cos\theta + 2C_4\,\cos^3\theta\right)}{\left(K_{11}\,\sin^2\theta_i^k + K_{33}\,\cos^2\theta_i^k\right)}\right] \quad \text{at} z = 0 \tag{45}$$

$$\theta_i^{k+1} = \theta_{i-1}^k - h_L\left[\frac{\sin\theta_i^k\left(C\cos\theta + 2C_4\,\cos^3\theta\right)}{\left(K_{11}\,\sin^2\theta_i^k + K_{33}\,\cos^2\theta_i^k\right)}\right] \quad at z = h_c \tag{46}$$

Note that the five-point difference method of solution reduces to a three-point difference method of solution near the boundaries. In order to obtain a faster convergent method, the relaxation factor is optimized numerically. The convergence of the five-point difference method of solution is faster than that of the corresponding three-point difference method of solution.

The external electric field is dissipated through an inverted microstrip line (MSL), which can control the orientation angle of LC molecules. In the case of molecules perpendicular to the electric field directions, it means no biased voltage (V = 0 V) is applied to IMSL. When the electric field exceeds the Freedericksz threshold voltage (V = 3.3 V), the LC molecules begin to reorient toward the electric field direction. By increasing the biased voltage until reaching the saturation point (V = 20 V), it means the LC molecules become parallel to the electric field directions inside the LC cell. The directors of molecules can be oriented in such a way as to form an angle $\theta(z)$ with respect to the z-direction of the cell. The orientation angle of molecules depends on the applied voltage as well as their position in the LC cell. **Figure 5** illustrates the orientation angle as a function of molecule position in the LC cell ($0 \le z \le h_c$) for different values of the applied voltage. As the applied voltage increases, the directors of NLC molecules fully align themselves with the z-direction.

It can be seen that the maximum value of the director's orientation angle in the middle of the LC cell ($z = hc/2$) is shown in the plot for the maximum director angles (θ_m) as a function of applied voltage in **Figure 6**, which is a more accurate approach

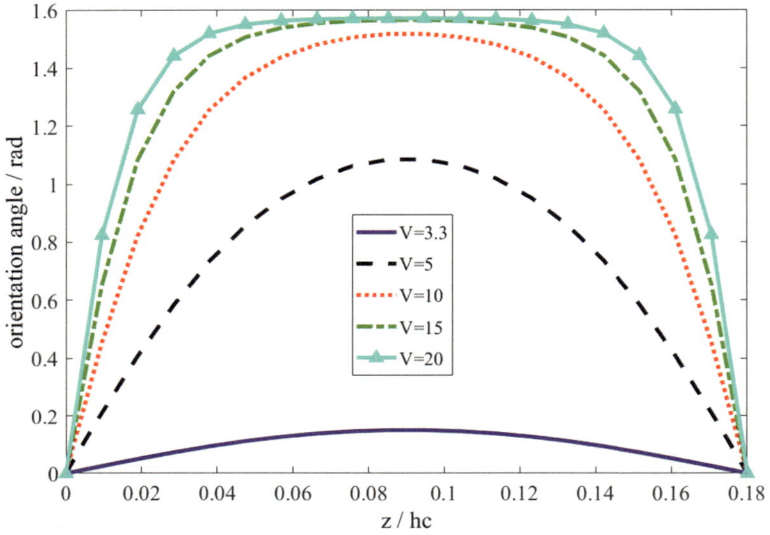

Figure 5.
The orientation angle of molecules versus their position inside the LC cell for different applied voltages.

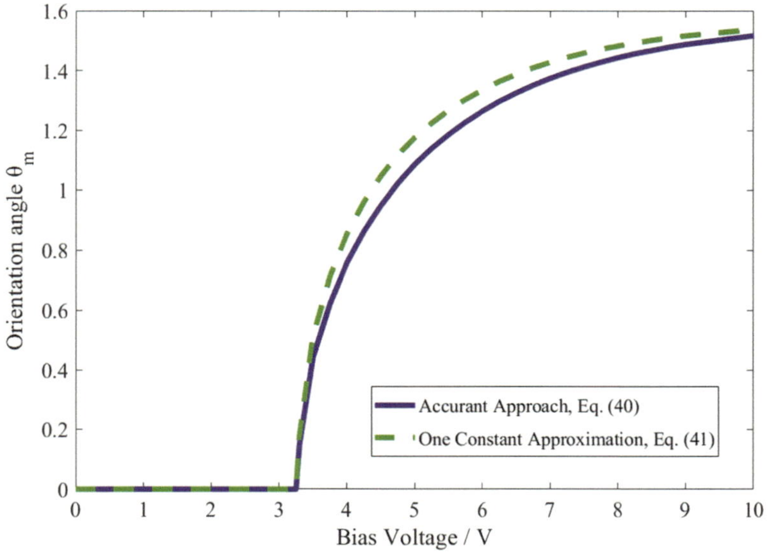

Figure 6.
The maximum director angle of LC molecules versus applied voltage.

derived from Eq. (40) than the one-constant approximation ($K_{11} = K_{22} = K_{33}$) in Eq. (41). The one-constant approximation assumes that the splay, twist, and bend deformation elastic constants are equal to simplify handling the numerical solutions for the differential equation of the LC orientation angle.

4.4 Response time for LC

It is important to define the generalized relations between LC and the behavior of the tuning speed or the switching times. In order to achieve the necessary relations, Eq. (40) is rewritten by replacing its zero right side with the retarding torque, which illustrates the damping of the spring-mass system as:

$$\frac{1}{2}\left(\frac{V}{V_{th}}\right)^2\left(\frac{\pi}{h_c}\right)^2\frac{\sin 2\theta}{(K_{11}\ \sin^2\theta+K_{33}\ \cos^2\theta)}-\frac{1}{2}\frac{(K_{11}-K_{33})\ \sin 2\theta}{(K_{11}\ \sin^2\theta+K_{33}\ \cos^2\theta)}\left(\frac{\partial\theta}{\partial z}\right)^2-\frac{\partial^2\theta}{\partial z^2} = \frac{\gamma_1}{K_{eff}}\cdot\frac{\partial\theta}{\partial t} \tag{47}$$

where γ_1 – Rotational viscosity of LC,
K_{eff} – the effective elastic constant.

Equation (47) is commonly called the Ericksen-Leslie equation. Due to the molecules anchoring at the surfaces, the LC molecules will be reoriented back to their normal position (no-order positional state) when the applied voltage is released. The mechanism of the response time (t_0) can commonly be derived as:

$$t_0 = \frac{\gamma_1}{K_{eff}}\cdot\frac{{h_c}^2}{\pi^2} \tag{48}$$

The K_{eff} is mostly given for the splay distortion K_{11} (also instead of the twist distortion K_{22}). On the other hand, the response time can be defined according to:

$$\theta_m(t) = Ce^{\left(\frac{t}{t_0}\right)} \tag{49}$$

where $\theta_m(t)$ – the angle in the middle of the LC cell ($z=hc/2$),
C – the arbitrary constant.

According to the characteristics of t_0, the rising time (t_{rising}) (the switch-on time) can be represented as the time taken for the rising tuning voltage to go from 10% to 90% of the delay time. Meanwhile, the decay time (t_{decay}) (the switch-off time) can be represented as the time taken for the voltage to release from 90% to 10% of the delay time when it returns to zero. The rising and decay times can be obtained as [25]:

$$t_{rise} = \frac{t_0}{\left|\frac{V}{V_{th}}-1\right|} \tag{50}$$

$$t_{decay} = \frac{t_0}{\left|\frac{V_{rem}}{V_{th}}-1\right|} \tag{51}$$

where V_{rem} – The remaining voltage when the applied voltage is removed.

The times Eqs. (50) and (51) can be represented as the maximum tuning speed. It will be a faster tuning process when a higher switching-on voltage is used. Meanwhile, it will be faster when the remaining voltage is lower.

5. IMSL phase shifter based on LC

5.1 Preliminary analysis of phase shifter

In general, the planar and waveguide structures can be represented as existing microwave topologies. The planar structure topology is a 2D circuit that can be used with printed TL technology for fabrication [23]. **Figure** 7 shows some examples of planar topologies, such as MSL, coplanar waveguide (CPW), stripline, and grounded coplanar waveguide (GCPW).

On the other hand, the waveguide topology is considered a 3D structure that uses a single conductor, which cannot support the transverse electromagnetic mode (TEM), but it can support the transverse electric (TE) and transverse magnetic (TM) modes, which can be propagated above a cut-off frequency only. The tunable microwave structure is achieved when the LC is replaced with the dielectric material in the planar or waveguide structures, as shown in **Figure 8**.

Moreover, the NLC cavity is dug into the LC spacer and closed from the top and bottom by the top substrate and ground plane, respectively, to prevent the liquid crystal from leaking outside the device, which makes the LC spacer house the NLC as well, which is sufficient for allowing the long LC molecules to orient in parallel toward the electric field. The bottom substrate is added to support the structure.

The change in the permittivity of the LC will be adjusted to the electric length of the TL, which utilizes the tunable delay line concept. Therefore, the phase shift (ϕ) equation depends on the permittivity of the LC for a TL with a phase constant (β) as [30]:

$$\phi = \beta \ell \tag{52}$$

where ℓ – the physical length of TL.

Based on LC properties, the TL has two extreme values of phase constant, which are $\beta_{\perp}$ and $\beta_{\parallel}$, with phase shifts $\phi_{\perp}$ and $\phi_{\parallel}$. As a result, the maximum differential phase shift can be defined for a phase shifter as:

$$\Delta\phi_{max} = \phi_{\perp} - \phi_{\parallel} = \left(\beta_{\perp} - \beta_{\parallel}\right)\ell \tag{53}$$

On the other hand, the differential phase shift for the IMSL phase shifter can be rewritten as [31]:

$$\Delta\phi_{\max} = \frac{2\,\pi\, f\, \ell}{c_0} \cdot \Delta n_{\text{eff}} \tag{54}$$

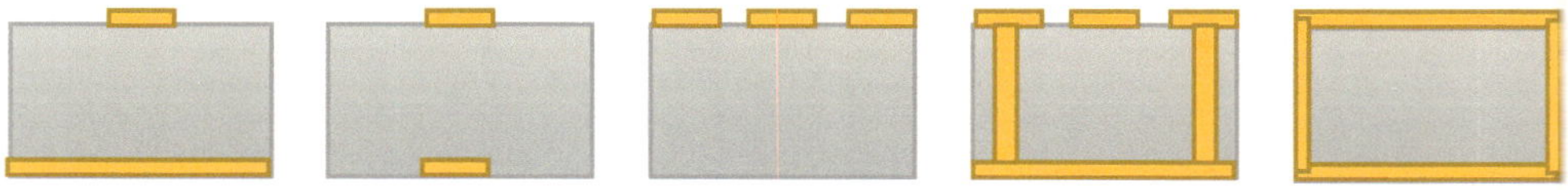

Figure 7.
The commonly used planar topologies.

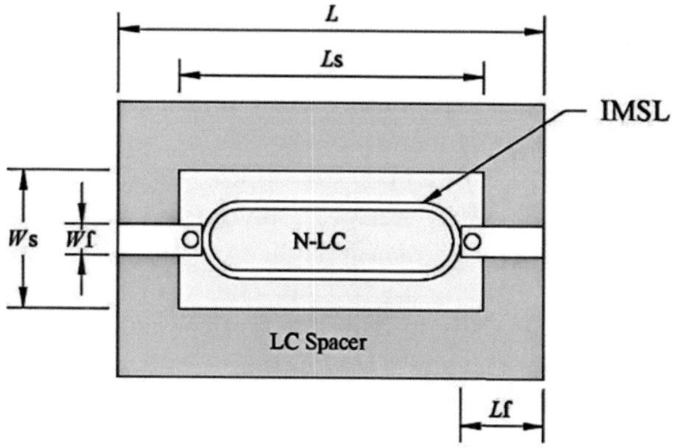

a) Top view

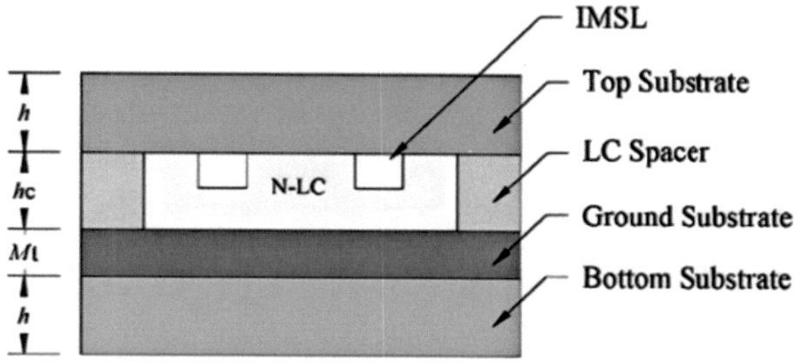

b) Cross-section view

Figure 8.
Cross-sectional view of double-IMSL phase shifter structure.

where $\Delta n_{\text{eff}} = \sqrt{\varepsilon_{\text{r}\|}} - \sqrt{\varepsilon_{\text{r}\perp}}$.

According to the properties of LC's loss tangent, the figure of merit (FoM) can be defined for a tunable phase shifter as [31]:

$$\text{FoM} = \frac{Vf_{\text{max}}}{IL_{\text{max}}} \tag{55}$$

Typically, the FoM has a range from 10–110°/dB for planar structures, which depends on the quality of the LC, other materials, and operating frequency. A higher FoM can be obtained with waveguide structures. Furthermore, the TL of these structures can end up with open or short circuits.

5.2 Design considerations of IMSL phase shifter

The novel IMSL phase shifter is constructed with three parts: the top substrate, the LC spacer, and the bottom substrate, as shown in **Figure 8**. The Rogers RT/Duroid 5880 dielectric material (with permittivity (ε_r) = 2.2 and dissipation factor ($\tan\delta$) = 0.0009) is used with h = 0.254 mm thickness for the top and bottom substrates. The LC spacer thickness, h_c, is 0.184 mm (which is h = 0.254 mm thickness after removing 1oz copper thickness).

An IMSL is one of the TL topologies used for obtaining a phase shift. It is located inside the NLC cell and placed on the bottom layer of the top substrate. The external electric field is applied through the IMSL TL to tune the LC molecules' position. The total length (L) of the phase shifter device is 28 mm. Meanwhile, the length of the NLC cell (L_s) is 20 mm, which has been chosen to be approximately equal to λ, and the width *(Ws)* is 8 mm. The NLC cell is surrounded by sidewalls of 4 mm, referred to as the LC spacer, which is located between the top substrate and the ground plane, as illustrated in **Figure 8b**. Moreover, the NLC cavity is dug into the LC spacer and closed from the top and bottom by the top substrate and ground plane, respectively, to prevent the liquid from leaking outside the device. This design makes the LC spacer house the NLC with a thickness *(hc)* of 0.184 mm, which is sufficient to allow the long LC molecules to orient parallel to the electric field. The bottom substrate is added to support the structure. On the other hand, the MSL, IMSL, and ground plane are copper-engraved with a thickness (Mt) of 0.035 mm.

Furthermore, the IMSL phase shifter is applied to investigate the tunable dielectric characteristics of NLC (GT3-23001) with perpendicular permittivity ($\varepsilon_\perp$ = 2.47, tanδ = 0.02) and parallel permittivity ($\varepsilon_\parallel$ = 3.25, tanδ = 0.015), so that the external electric field controls the positional order and orientation angle of LC molecules. The length of the IMSL TL is (Ls) = 20 mm, whose terminals are connected between two segments of MSLs, each having an L_f = 5 mm length and Wf = 2.5 mm estimated width of a 50 Ω microstrip feedline, while the IMSL width is 2.5 mm to obtain better orientation efficiency for LC molecules. Based on the cross-section of the phase shifter structure, the total width of the device is 15 mm, which makes it suitable and easy to incorporate into various microwave antennas and communication systems.

6. Analysis of IMSL phase shifter

6.1 Simulation results

The double (novel) IMSL phase shifter offers higher performance than a single (common) IMSL phase shifter. The scattering parameters of both (double-IMSL and single-IMSL) phase shifters are given in **Figures 9** and **10**, which show the results for the input reflectance ($S_{1,1}$) and the transmittance ($S_{2,1}$), respectively, under the control of an applied voltage ranging from 0 V to 20 V.

The return loss for the double-IMSL phase shifter is determined to be more than - 10 dB, with a frequency shift from 10.57 GHz to 8.86 GHz, while the single-IMSL phase shifter is from 11.1 GHz to 10.1 GHz. Meanwhile, the insertion loss changes from 0.56 dB to 1.06 dB for the double-IMSL phase shifter, while it changes from 0.8 dB to 1.9 dB for the single-IMSL phase shifter within the whole analyzed frequency range. Note that the single (common) IMSL phase shifter has the same dimensions as the double (novel) IMSL phase shifter.

Figure 11a shows the tunable transmittance phase versus the applied voltage for choosing the thickness of the double-IMSL phase shifter. The differential phase shift increased steadily up to 57°. On the other hand, this type of phase shifter has low tuning efficiency because of its low conductor loss. Therefore, the tuning efficiency practically depends on the properties of the top substrate, such as its thickness, as shown in **Figure 11b**, which illustrates the effect of top substrate thickness on the

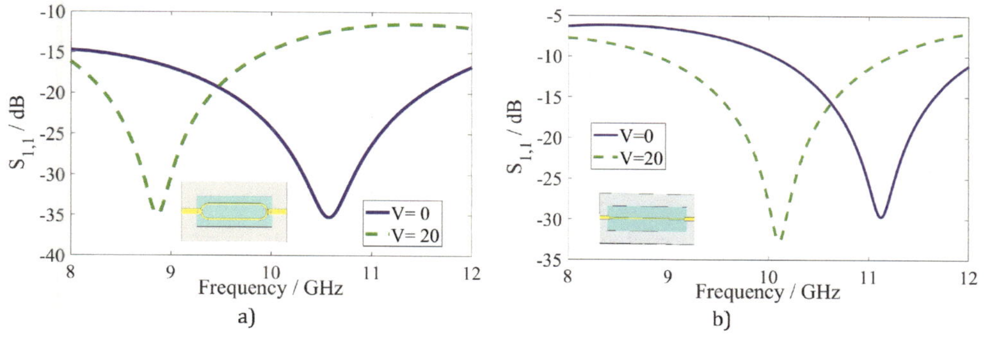

Figure 9.
The return loss $S_{1,1}$ *at 0V to 20V voltage.*

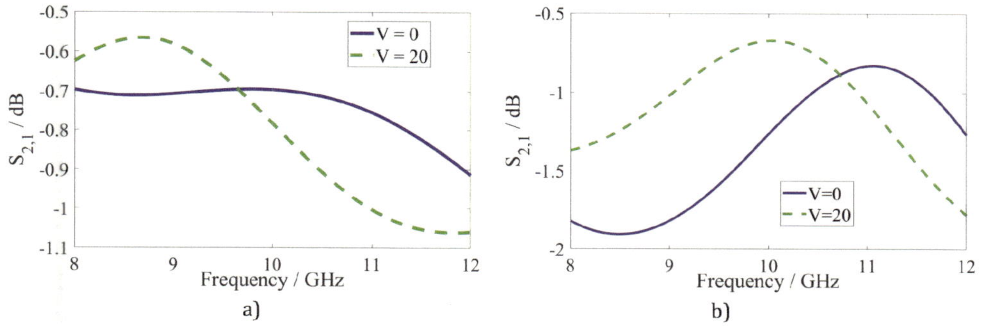

Figure 10.
The insertion loss $S_{2,1}$ *at 0V to 20V voltage.*

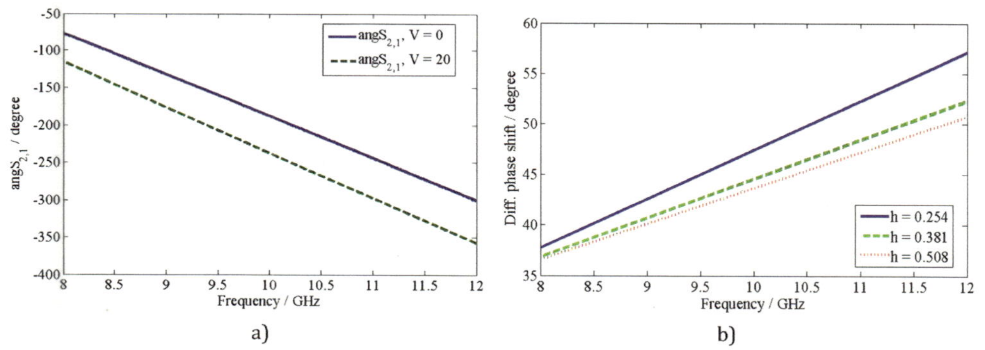

Figure 11.
(a) The tunable transmittance phase vs. frequency for different voltages. (b) The effect of top substrate thickness standard of Rogers RT5880 for different transmittance phase shifts.

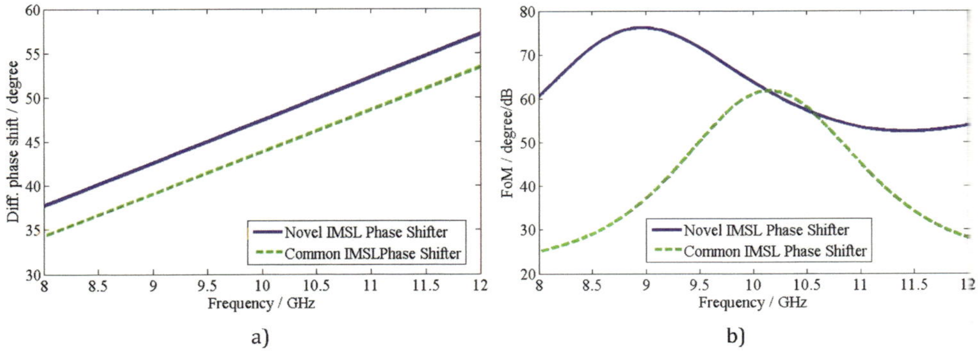

Figure 12.
(a) The effect of IMSL for different transmittance phase s, (b) The FOM vs. frequency for double- and single-IMSL phase shifter.

differential phase shift (Δϕ). The clear effect of the Rogers standard thickness for the top substrate is on the tunability of the device, where the tunability decreases with increasing the thickness of the top substrate, as well as increasing its permittivity.

The tuning efficiency improved with a thinner top substrate, whereas the stability of the device and uniformity of the LC layer cannot be achieved with the thinner substrate. As well as the geometry of the IMSL, the tuning efficiency is also affected, as seen in **Figure 12a**. The double-IMSL phase shifter device has chosen a 0.254 mm thickness for the top substrate, which can improve the tuning efficiency and its stability. Meanwhile, the FoM shows higher values for the double-IMSL topology (FoM is up to 76.15°/dB) than the single-IMSL topology (FoM is up to 61.76°/dB), as shown in **Figure 12b**.

6.2 Measurement results

A vector network analyzer (VNA), calibrator, and biasing network are used during the measurement analysis. **Figure 13** illustrates the measurement setup that is used for the IMSL phase shifter. The VNA is connected to the IMSL phase shifter through a biasing network, which connects the VNA signal and the calibrator signal. The biasing network sums the RF signal and the bias voltage to feed the IMSL phase shifter through the port connector. The calibrator is set up to provide a DC voltage from 0 V to 20 V. The phase shifter's layers are stuck together using a gluey layer, which is sealed with silicon glue to prevent the LC from leaking out. Each port is equipped with an SMA connector (sub-miniature version A). The NLC is injected into the cavity of the LC spacer layer. Note that the voltage source ground is connected to the ground plane of the phase shifter structure.

After processing the structure's measurement, the phase shifter structure is injected with NLC and measured with the NLC-filled device. The antenna feeds have an input impedance close to a 50-ohm match at the desired frequency. The S-parameter shows good results for the measurement process, as seen in **Figure 14a**. When the biasing voltage from 0 V to 20 V is applied, a continuous tuning of the S-parameters ranges from 10.51 GHz to 8.89 GHz (up to 1.62 GHz). The $S_{1,1}$ parameter showed a slightly different value for the simulated and measured results because of SMA welding, which caused a slight impedance mismatch, a gluey layer

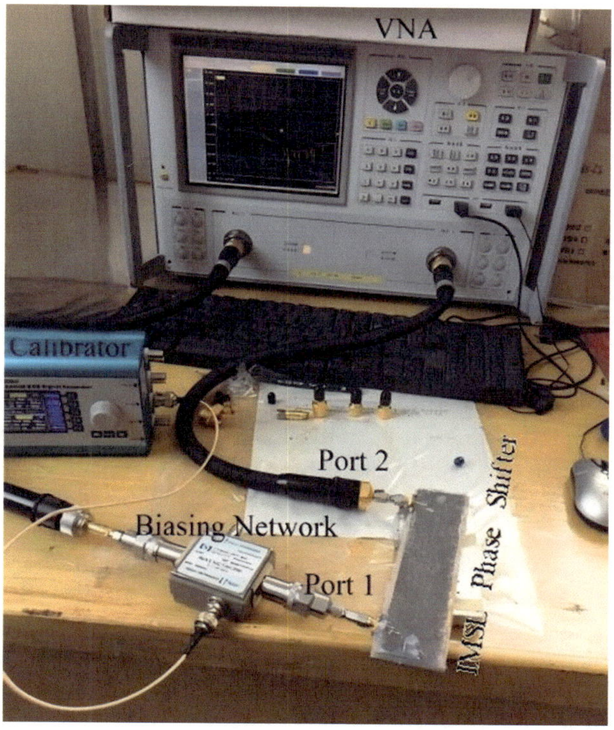

Figure 13.
The measurement setup for the IMSL phase shifter.

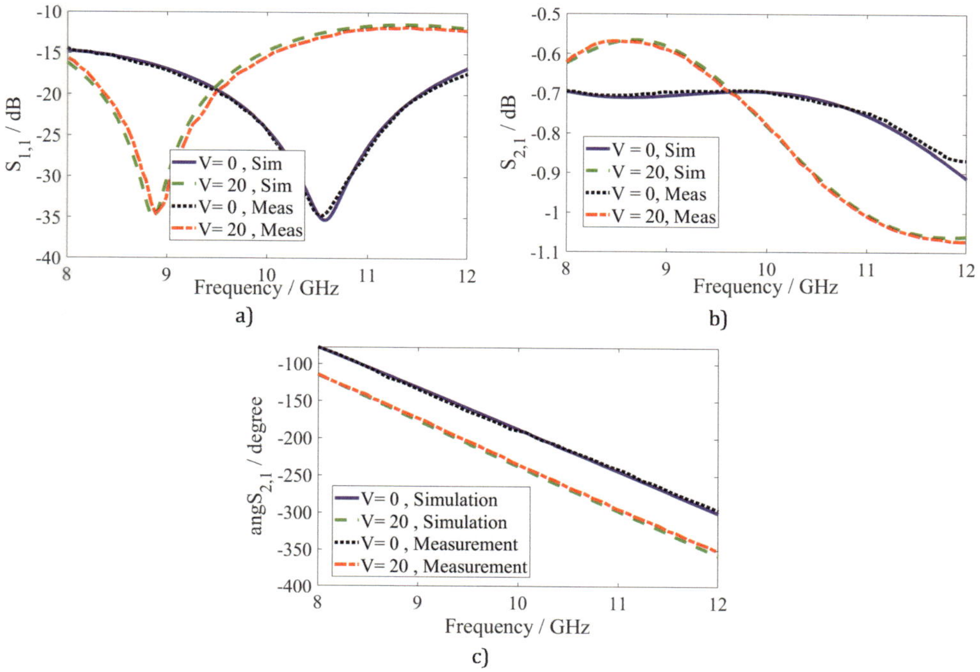

Figure 14.
Measured and simulated results: (a) S1,1, (b) S2,1, (c) ang S2,1.

that stuck the structure's layers, and some NLC molecules that may not have been fully oriented. A slight difference in the $S_{1,1}$ parameter (return loss) between the simulated and measured results is considered an acceptable value. Meanwhile, the $S2,1$ measured results show agreement with the simulation results, as illustrated in **Figure 14b**. Furthermore, the differential phase has also been measured, as seen in **Figure 14c**, which is tuned steadily up to 53°.

7. Conclusion

This chapter presents how to design and analyze an IMSL phase shifter device based on a tunable LC with a typical standard thickness of dielectric material. The dielectric characteristics of this tunable device in the X-band frequency range were measured with a double IMSL. This double-IMSL topology showed high performance in terms of tuning efficiency and stability. The dielectric permittivity properties and the loss tangent for NLC were investigated with GT3 23001 LC, which has the perpendicular permittivity ($\varepsilon\perp$ = 2.47, tanδ = 0.02) and the parallel permittivity ($\varepsilon\|$ = 3.25, tanδ = 0.015). To understand the theory of LC, the orientation angle is derived and solved numerically for the electric field using finite-difference schemes. A solution for the differential equation of the orientation angle is obtained with more accuracy than one approximation constant ($K_{11} = K_{22} = K_{33}$). The convergence of the three-point difference method of solution took 557s to complete the numerical solution. As a result, when the applied DC voltage increased from 0 V to 20 V, the differential phase shift of the device tuned up to 57° with low insertion loss from 0.56 dB to 1.06 dB, and a frequency shift from 10.57 GHz to 8.86 GHz. Meanwhile, the FoM value is obtained up to 76.15°/dB. The double-IMSL phase shifter shows higher performance than a single-IMSL phase shifter. The measured results showed a slightly different value from the simulated results due to SMA welding, which caused a slight impedance mismatch, a gluey layer that stuck the structure's layers, and some NLC molecules that may not have been fully oriented. This tunable phase shifter can be a suitable device and is easy to integrate into microwave antennas or communication systems for achieving phase shift, which requires intensive research to guarantee further advancements.

Acknowledgments

My Appreciation is extended to Harbin University of Science and Technology for providing the facilities and resources necessary for conducting the work presented in this chapter. My Appreciation is also extended to Heilongjiang Basic Research Funds for Undergraduate Universities (Grant No. 2023-KYYWF-0108) for support. I would like to thank my colleagues for their constructive discussions and valuable feedback, which contributed to the refinement of this chapter.

Conflict of Interest

The authors declare no conflict of interest.

DOI: http://dx.doi.org/10.5772/intechopen.1013776

Author details

Odai Hassan Raheem Al Soad[1]*, Wan Chen[2] and Zaid Shakir Kadhim Al-Shammari[3]

1 School of Measurement–Control Technology and Communications Engineering, Harbin University of Science and Technology, Harbin, China

2 College of Information and Communication Engineering, Harbin Engineering University, Harbin, China

3 College of Engineering, Al-Karkh University of Sciences, Baghdad, Iraq

*Address all correspondence to: odai@hrbust.edu.cn

References

[1] Zhang Y, Zhu L, Xiao S, Tang W, Eldar YC, et al. Movable Antenna–Aided hybrid beamforming for multi-user communications. IEEE Transactions on Vehicular Technology. 2025:1–6. DOI: 10.1109/TVT.2025.3533078

[2] Jaiswal A, Dey S, Abegaonkar MP, Koul SK. A 77-GHz polarization-agile microelectromechanical system antenna. Microwave and Optical Technology Letters. 2020;**62**(6):2300–2306. DOI: 10.1002/mop.32292

[3] Cao J, Jx L, XW L, Li SK, Xue XY. Effect of the distribution state of transition phase on the mechanical properties and failure mechanisms of the W-Mo-Cu alloy by tuning elements content. Journal of Alloys and Compounds. 2020;827. DOI: 10.1016/j.jallcom.2020.154333

[4] Tan MJH, Patel SK, Chiu J, Zheng ZT, Odom TW. Liquid lasing from solutions of Ligand-Engineered semiconductor nanocrystals. Journal of Chemical Physics. 2024;**160**(15). DOI: 10.1063/5.0201731

[5] Mu DS, Lakshmi T, Kumaresan D, Prasad RK, Sivakumar M. Hydrothermal synthesis, phase control, and dielectric analysis of niobium-doped barium strontium titanate. Journal of Materials Science: Materials in Electronics. 2025;**36**(14). DOI: 10.1007/s10854-025-14873-1

[6] Wang MY, Lee JS, Aggarwal S, Farmakidis N, He YH, Cheng TS, et al. Varifocal metalens using tunable and Ultralow-loss dielectrics. Advanced Science. 2023;**10**(6). DOI: 10.1002/advs.202204899

[7] Fu J, Raheem OH. A novel IMSL tunable phase shifter for HMSIW-LWA-fed rectangular patches based on nematic liquid crystal. Applied Physics A: Materials Science & Processing. 2017;**123**(7). DOI: 10.1007/s00339-017-1081-4

[8] Al Soad OHR, Zhou CF, Fu JH, Wu Q, IEEE. An Electrically Tunable Leaky-wave antenna based on slotted metagratings using liquid Crystal. Proceedings of the 2021 Cross Strait Radio Science and Wireless Technology Conference (Csrswtc). 2021:123–125. DOI: 10.1109/CSRSWTC52801.2021.9631644

[9] Kaesser T, Fritzsch C, Franz M. Tunable RF filters based on liquid crystal for space applications. Crystals. 2020;**10**(6):455. DOI: 10.3390/cryst10060455

[10] Chen P, Wang D, Lh W, Liu L, Gan ZS. Liquid crystal-based reconfigurable antenna for 5G millimeter-wave. Scientific Reports. 2024;**14**(1). DOI: 10.1038/s41598-024-67714-2

[11] Al Soad OHR, Fu JH, Wu Q. A tunable ǂ-shaped microstrip leaky wave antenna based on liquid crystal. International Journal of Communication Systems. 2022;**35**(10). DOI: 10.1002/dac.5175

[12] Krakhalev MN, Prishchepa OO, Sutormin VS, Zyryanov VY. Polymer dispersed nematic liquid crystal films with conical boundary conditions for electrically controllable polarizers. Optical Materials. 2019;**89**:1–4. DOI: 10.1016/j.optmat.2019.01.004

[13] Pouria R, Edalatpour S. Tuning the spectrum of Near-Field radiative heat transfer using Mie resonance based metamaterials. Physical Review B. 2024;**109**(4). DOI: 10.1103/PhysRevB.109.045407

[14] Behera BR, Alsharif MH, Jahid A. Investigation of a circularly polarized metasurface antenna for hybrid wireless applications. Micromachines. 2023;**14**(12):2172. DOI: 10.3390/mi14122172

[15] Deng LM, Lin SS, Pham TD, Chen YH. Optimized Domain-Engineered lithium niobate for Electro-Optic spectral tuning in a Multi-wavelength optical parametric oscillator. Optics Communications. 2025;574. DOI: 10.1016/j.optcom.2024.131129

[16] Coles HJ, Pivnenko MN. Liquid crystal 'Blue Phases' with a wide temperature range. Nature. 2005;**436**(7053):997–1000. DOI: 10.1038/nature03932

[17] Yang DK, Wu ST. Fundamentals of Liquid Crystal Devices. New Jersey: John Wiley & Sons; 2014. 48–127 p. DOI: 10.1002/9781118751992

[18] Gölden F. Liquid crystal based microwave components with fast response times: Material, technology, power handling capability, Vol. **1**. Technische Universität; 2010. 15–28 p. DOI: 10.26083/tuprints-00002203

[19] Gray GW, Vill V, Spiess HW, Demus D, Goodby JW. Physical Properties of Liquid Crystals. GmbH: John Wiley & Sons; 2009. 6–30 p. DOI: 10.1002/9783527613946

[20] Elliott RS. Electromagnetics: History, Theory and Applications. New York: IEEE Computer Society Press; 1993. 53–69 p

[21] Pozar DM. Microwave Engineering. New York: John Wiley & Sons; 2009. 114–131 p

[22] Yang D-K. Fundamentals of Liquid Crystal Devices. New York: John Wiley & Sons; 2014. 7–51 p

[23] Karabey OH. Electronic Beam Steering and Polarization Agile Planar Antennas in Liquid Crystal Technology. Switzerland: Springer Science & Business Media; 2013. 10–23 p. DOI: 10.1007/978-3-319-01424-1

[24] Karabey OH, Bildik S, Fritzsch C, Strunck S, Gaebler A, Jakoby R, et al. Liquid crystal based reconfigurable Antenna arrays. Proceedings of the 32nd ESA Antenna workshop; 2010. p. 159–168

[25] Khoo I-C, Wu S-T. Optics and Nonlinear Optics of Liquid Crystals. New Jersey: World Scientific; 1993. 51–75 p. DOI: 10.1142/1630

[26] Willman E, Fernández FA, James R, Day SE. Modeling of weak anisotropic anchoring of nematic liquid crystals in the Landau–de Gennes theory. IEEE Transactions on Electron Devices. 2007;**54**(10):2630–2637. DOI: 10.1109/TED.2007.904369

[27] Prost J. The Physics of Liquid Crystals. London : Oxford university press; 1995. 42–89 p. DOI: 10.1093/oso/9780198520245.001.0001

[28] Blinov LM, Chigrinov V. Electrooptic Effects in Liquid Crystal Materials. New York: Springer Science & Business Media; 2012. 27–50 p. DOI: 10.1007/978-1-4612-2692-5

[29] Priestly E. Introduction to Liquid Crystals. New York: Springer Science &

Business Media; 2012. 61–69 p. DOI: 10 .1007/978-1-4684-2175-0

[30] Misra D. Radio-frequency and Microwave Communication Circuits: Analysis and Design. New York: John Wiley & Sons; 2004. 131–146 p

[31] Muller S, Scheele P, Weil C, Wittek M, Hock C, Jakoby R, editors. Tunable passive phase shifter for microwave applications using highly anisotropic liquid crystals. Microwave Symposium Digest, 2004 IEEE MTT-S International; IEEE; 2004. DOI: 10.1109/ MWSYM.2004.1339190

Chapter 6

Enhancing Wireless Systems Performance Using Antenna Distance-based Modulation Approach

Somayeh Mohammady

Abstract

In modern wireless communications, efficient spectrum utilization is critical to achieving high data throughput while preserving signal reliability. As the distance between transmitter and receiver grows, signal attenuation and noise increase, reducing the signal-to-noise ratio (SNR) and impairing system performance. This paper presents a distance-aware modulation adaptation strategy that dynamically selects optimal modulation schemes—ranging from BPSK and QPSK to high-order QAM—based on real-time SNR measurements. We explore the theoretical and practical foundations of SNR, data rate, and bit error rate (BER), applying the Free-Space Path Loss (FSPL) model and the Shannon Capacity Theorem to guide our methodology. Extensive MATLAB simulations compare the performance of fixed and adaptive modulation schemes across varying distances. Results reveal that the proposed approach significantly improves spectral efficiency and data throughput while maintaining robust communication, particularly in the context of 5G and emerging wireless technologies.

The results indicate that higher-order modulation schemes, such as 32-QAM and 32-QPSK, experience increased BER at a given distance due to reduced noise tolerance, while lower-order schemes like BPSK and 4-QAM offer improved reliability over longer distances by operating at lower SNR levels. As distance increases, SNR degrades logarithmically and BER rises, necessitating adaptive modulation. High-order modulation is suitable at short distances with high SNR, whereas increasing distance requires a transition to more robust lower-order schemes, beyond which only low-data-rate modulation can sustain reliable communication. This makes distance-based adaptive modulation a vital enabler of resilient, high-performance next-generation networks.

Keywords: antenna, BER, BPSK, modulation scheme, SNR, QAM, QPSK

1. Introduction

The performance of wireless communication systems is heavily influenced by the Signal-to-Noise Ratio (SNR), which in general is known to vary as a function of distance from the transmitter. As wireless technologies evolve to meet the

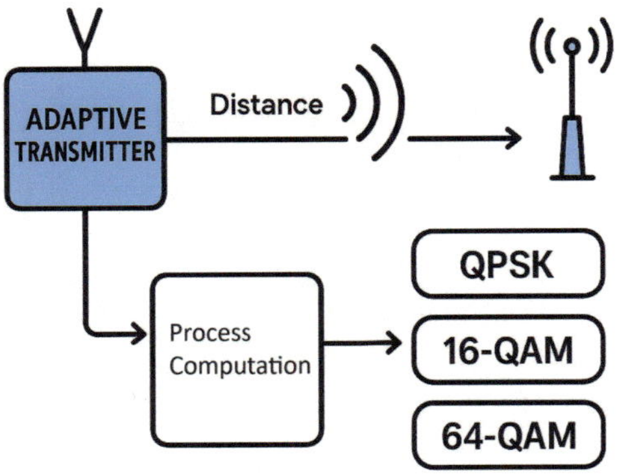

Figure 1.
The adaptive modulation system based on distance from the antenna.

demands of higher data rates, ultra-low latency, and massive device connectivity, SNR has become a key metric influencing system design and operation. The idea is illustrated in **Figure 1.** and its importance is underscored across several dimensions:

- **Communication Quality and Reliability:** SNR quantifies the strength of a signal relative to background noise. A high SNR ensures accurate signal detection and decoding, reducing bit error rates and enhancing the reliability of data transmission. In critical applications such as 5G, autonomous systems, and industrial IoT, maintaining a strong SNR is essential for stable and secure communication [1, 2].

- **Achievable Data Rates:** According to the Shannon Capacity Theorem,

$$C = B\log_2(1 + \text{SNR}),$$

where C is the channel capacity and B is the bandwidth [3]. This relationship shows that the data rate achievable over a channel increases logarithmically with SNR. High SNR enables higher-order modulation schemes, directly contributing to faster and more efficient communication [4].

- **Adaptive Modulation and Coding (AMC):** Contemporary systems dynamically adjust their modulation and coding schemes based on real-time SNR measurements. Higher SNR allows the use of complex modulations (e.g., 64-QAM, 256-QAM), increasing throughput, while lower SNR necessitates simpler, more robust schemes (e.g., BPSK, QPSK).

- **Spectral Efficiency and Network Capacity:** SNR directly impacts spectral efficiency—the amount of information transmitted per unit bandwidth. In high-density deployments such as urban 5G cells and smart city infrastructures, optimizing spectral efficiency is essential to support large numbers of simultaneous users [5].

- **Support for Emerging Applications:** Applications such as augmented/virtual reality (AR/VR), edge computing, and ultra-reliable low-latency communications (URLLC) demand high SNR to ensure seamless performance. Similarly, mmWave and satellite systems, which operate under challenging propagation conditions, require robust SNR management strategies.

- **Energy Efficiency:** Low SNR leads to increased retransmissions and heavier error correction, which in turn increases energy consumption. Systems that maintain high SNR can operate more efficiently, which is particularly beneficial for battery-powered IoT devices [6].

Given its central role in determining communication quality, data throughput, and energy efficiency, SNR serves not only as a performance metric but also as a guiding parameter in the design and optimization of future wireless systems.

Here, the methodology used to analyze the SNR vs. Distance is explained. It should be noted that the measurements are performed based on asumption of the Free-Space Path Loss (FSPL) model [7]. The FSPL is mathematically expressed as:

$$FSPL(dB) = 20\log_{10}(d) + 20\log_{10}(f) + 20\log_{10}\left(\frac{4\pi}{c}\right) \quad (1)$$

where d is the transmitter-receiver distance (m), f is the signal frequency (Hz), and c is the speed of light (3×10^8 m/s). For practical calculations:

$$FSPL(dB) = 32.4 + 20\log_{10}(d) + 20\log_{10}(f) \quad (2)$$

when d is in kilometers and f is in MHz.

1.1 Path loss and distance

FSPL increases logarithmically with distance. The comparisions are carried out assuming a clear Line of sight (LOS) (No massive onstacle between Transmitter (Tx), and Receiver (Rx) as shown in **Figure 2.** Doubling the distance increases FSPL by 6 dB, reducing received power by a factor of four [8, 9].

Figure 2.
Clear LOS (no object) between the transmitter and the receiver.

1.2 Path loss and frequency

Higher frequencies suffer from greater FSPL, making low-frequency bands (e.g., sub-6 GHz) more suitable for long-range applications, whereas mmWave (28 GHz, 60 GHz) requires small cells and beamforming [10].

1.3 Impact of FSPL on wireless networks, i.e. 5G and mmwave communication

Millimeter-wave frequencies (28 GHz, 39 GHz) face high FSPL, limiting their range. Beamforming and Massive MIMO mitigate these effects [11].

1.4 FSPL in iot and low-power networks

Long-range IoT networks (e.g., LoRaWAN, Sigfox) rely on low-frequency bands (e.g., 868 MHz, 915 MHz) to minimize FSPL and extend coverage [12].

1.5 FSPL in simulations

A MATLAB-based simulation was conducted to analyze FSPL at different frequencies and distances. The results are summarized in **Table 1**.

1.6 Limitations of FSPL model

FSPL assumes free-space conditions and ignores:

- Multipath fading (reflections, diffractions)
- Obstacles (walls, buildings, trees)
- Atmospheric absorption at high frequencies

For real-world scenarios, extended models like Okumura-Hata and COST-231 are used.

1.7 Impact of SNR on data rate in wireless communication

Signal-to-Noise Ratio (SNR) is a fundamental parameter that dictates the performance of wireless communication systems. It directly affects the achievable

Frequency (GHz)	Distance (m)	FSPL (dB)
2.4	100	80.1
5	100	86.1
28	100	104.2
60	100	112.0

Table 1.
FSPL at different frequencies and distances.

data rate, error performance, and spectral efficiency. The impact of SNR on data rate can be analyzed through the Shannon Capacity theorem, modulation schemes, and practical system considerations.

1.7.1 Shannon capacity and SNR

The relationship between Shannon Capacity and Signal-to-Noise Ratio (SNR) is fundamental to understanding the performance limits of wireless communication systems. According to the Shannon–Hartley theorem, the channel capacity C (in bits per second) is given by $C = B\log_2(1 + \text{SNR})$, where B is the channel bandwidth. This logarithmic relationship highlights that increases in SNR lead to proportional improvements in capacity at low-to-moderate SNR levels, enabling higher data rates and more efficient spectrum utilization. However, at very high SNR values, the capacity gain per unit increase in SNR diminishes, indicating a point of diminishing returns. This interplay is crucial in system design, as it dictates the optimal trade-off between transmitter power, bandwidth allocation, and modulation scheme selection. In practical systems, especially in 5G and beyond, maintaining an adequate SNR is essential for enabling higher-order modulation schemes, achieving low bit error rates (BER), and ensuring robust, high-throughput communication across varying propagation environments [1–3].

The theoretical upper bound on data rate is given by the Shannon-Hartley theorem:

$$C = B\log_2(1 + \text{SNR}) \tag{3}$$

where C is the channel capacity (bits per second), B is the channel bandwidth (Hz), and SNR is the signal-to-noise ratio expressed as a power ratio [13, 14].

From Equation (1), key observations include:

- Higher SNR results in a higher channel capacity, allowing more information to be transmitted reliably.
- The relationship is logarithmic, meaning that at very high SNR, further increases provide diminishing returns in capacity.
- For low-SNR scenarios, capacity increases approximately linearly with SNR.

1.7.2 Impact on modulation and coding

Different modulation schemes require specific SNR thresholds to maintain an acceptable bit error rate (BER). Higher-order modulations provide higher spectral efficiency but require better SNR conditions:

- Low SNR: Robust but low-efficiency schemes such as Binary Phase Shift Keying (BPSK) and Quadrature Phase Shift Keying (QPSK) are used.
- Moderate SNR: Higher-order modulations like 16-Quadrature Amplitude Modulation (16-QAM) are employed.
- High SNR: 64-QAM and 256-QAM offer higher data rates but require stronger signal quality [15, 16].

The minimum SNR requirement for an M-QAM system can be approximated as:

$$\text{SNR}_{\text{dB}} \approx 10\log_{10}\left(\frac{3M}{M-1}\right) \tag{4}$$

where M represents the modulation order. For example, 16-QAM requires about 12 dB SNR, while 256-QAM needs approximately 24 dB SNR for reliable communication.

1.7.3 Adaptive modulation and coding (amc)

Modern wireless systems (e.g., LTE, 5G, Wi–Fi) employ Adaptive Modulation and Coding (AMC) to dynamically adjust the modulation and coding scheme based on real-time SNR measurements. This ensures optimal data transmission:

- If SNR is high, the system selects higher-order modulation and a lower coding redundancy, maximizing throughput.
- If SNR degrades, a lower-order modulation and stronger error correction are used to maintain reliability [9, 17].

1.7.4 Practical considerations

Beyond theoretical limits, real-world wireless systems face several SNR-dependent challenges:

- **Interference and Fading**: Co-channel interference and multipath fading degrade SNR, requiring robust error correction techniques [3, 7].
- **Bandwidth-SNR Trade-off**: Increasing bandwidth can increase capacity, but also captures more noise, reducing effective SNR.
- **MIMO and Diversity**: Multiple-Input Multiple-Output (MIMO) and beamforming techniques help mitigate SNR degradation, improving spectral efficiency.

1.7.5 Summary

1.7.5.1 Impact of SNR on wireless system performance

Thus, SNR is a fundamental factor in determining achievable data rates, influencing modulation selection, coding strategies, and overall wireless system efficiency [11, 12]. **Table 2.** highlights low SNR and high SNR and related scenarios in terms of modulation and BER.

To address the challenges highlighted in the abstract and build upon the distance-based modulation framework proposed in the title, the remainder of this paper is organized as follows. Section II provides a comprehensive analysis of the Free-Space Path Loss (FSPL) model and its implications on signal degradation over distance. Section III delves into the mathematical relationship between Signal-to-Noise Ratio (SNR), modulation schemes, and achievable data rate, grounded in the Shannon

Factor	Low SNR Impact	High SNR Impact
Data Rate	Low	High
Modulation	BPSK, QPSK	16-QAM, 64-QAM, 256-QAM
Bit Error Rate	High	Low
Coding	Strong error correction	Minimal correction
Interference Sensitivity	Severe degradation	Minor degradation

Table 2.
Summarizes the key impacts of SNR on data rate and system performance.

Capacity Theorem. Section IV presents simulation results that illustrate how modulation schemes perform across varying distances under realistic wireless channel conditions. Section V introduces and evaluates a novel metric, the Modulation Adaptability Score (MAS), to quantify the suitability of modulation schemes as a function of distance. Finally, Section VI concludes the paper with a discussion of findings, implications for future wireless systems such as 5G and beyond, and potential directions for further research. This structured approach ensures a coherent narrative from theoretical foundations to practical validation, reinforcing the value of distance-aware modulation adaptation in enhancing wireless system performance [4, 18].

2. Mathematical analysis

The impact of Signal-to-Noise Ratio (SNR) on data rate in 5G communication is governed by two key formulations: the *Shannon Capacity Theorem* and practical adaptive modulation schemes.

2.1 Shannon capacity

The theoretical upper bound of channel capacity is given by:

$$C = B \log_2(1 + \text{SNR}) \tag{5}$$

where:

- C is the channel capacity in bits per second,
- B is the channel bandwidth in Hz (set to 100 MHz in our simulation),
- SNR is the signal-to-noise ratio in linear scale.

As sen on **Figure 3**, the *Shannon Capacity* is indicated as dashed black line as a theoretical limit, providing an ideal benchmark for comparison with practical modulation schemes [2, 19]. In real-world 5G systems, adaptive modulation is used to optimize data transmission. The spectral efficiency for an M-ary Quadrature Amplitude Modulation (M-QAM) scheme is given by:

$$R = B \log_2(M) \tag{6}$$

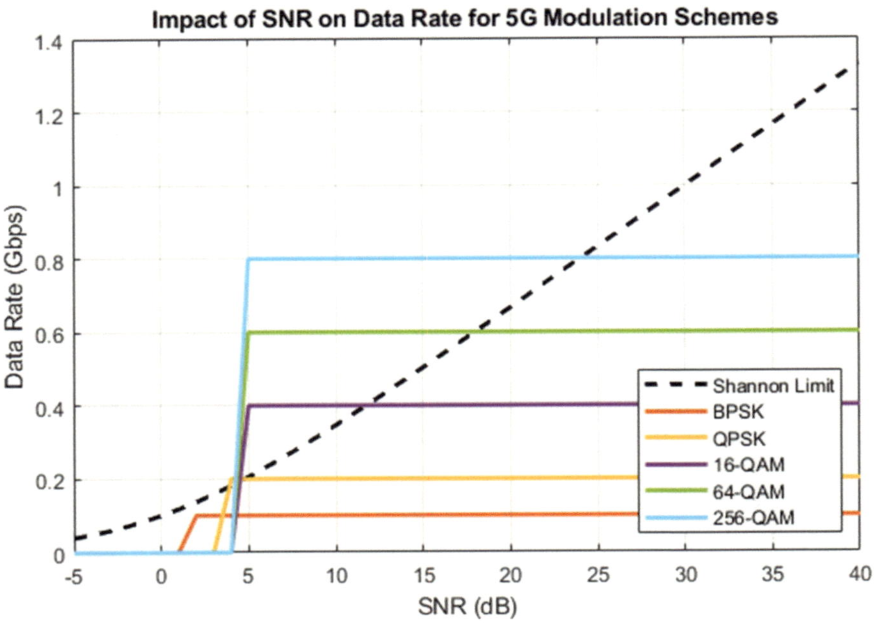

Figure 3.
SNR vs. Data rate for 5G modulation schemes.

where:

- M is the modulation order (e.g., BPSK: $M = 2$, QPSK: $M = 4$, etc.),
- R is the achievable data rate in bits per second.

However, each modulation scheme requires a minimum SNR threshold SNR_{min} to ensure reliable communication. The threshold is approximated as:

$$\text{SNR}_{\text{min}} \approx 10 \log_{10}\left(\frac{3(M-1)}{M}\right) \tag{7}$$

If the available SNR is below SNR_{min}, that modulation scheme is disabled to prevent high bit error rates. **Figure 3** visualizes the impact of SNR on data rate across different 5G modulation schemes.

- The **Shannon Capacity** (black dashed line) represents the theoretical upper bound.
- Each colored line represents a practical modulation scheme (BPSK, QPSK, 16-QAM, 64-QAM, 256-QAM).
- At low SNR values, only low-order modulation schemes (BPSK, QPSK) are usable, leading to lower data rates.
- As SNR increases, higher-order modulation schemes (16-QAM, 64-QAM, 256-QAM) become viable, increasing the data rate.

- The diminishing returns effect is visible: at very high SNR, even large increases in SNR result in only marginal data rate improvements.

2.2 Signal-to-noise ratio (SNR) in wireless communication

SNR quantifies the quality of a received signal relative to background noise, and it is defined as:

$$SNR = \frac{P_s}{P_n} \tag{8}$$

where P_s represents the received signal power, and P_n denotes the noise power [1, 20]. In practical wireless systems, SNR is often expressed in decibels (dB) for easier interpretation:

$$SNR_{dB} = 10\log_{10}\left(\frac{P_s}{P_n}\right) \tag{9}$$

A higher SNR indicates a stronger, more distinguishable signal relative to noise, whereas a lower SNR suggests a weaker, more corrupted signal [9, 17].

2.2.1 SNR dependency on distance

In free-space and real-world wireless channels, SNR decreases as the receiver moves farther from the transmitter due to path loss. This relationship is governed by the Free-Space Path Loss (FSPL) model:

$$P_r = P_t G_t G_r \left(\frac{\lambda}{4\pi d}\right)^2 \tag{10}$$

where:

- P_r is the received power,
- P_t is the transmitted power,
- G_t and G_r are the transmitter and receiver antenna gains, respectively,
- λ is the signal wavelength, and
- d is the distance from the transmitter.

Since noise power (P_n) remains relatively constant in most practical scenarios, SNR follows an inverse-square law with distance:

$$SNR(d) \propto \frac{1}{d^2} \tag{11}$$

which implies that as the distance doubles, the received SNR is reduced by approximately 6 dB.

2.3 Bit error rate (BER) in wireless communication

Bit Error Rate (BER) is a key performance metric that quantifies the reliability of a wireless communication system by measuring the proportion of incorrectly received bits over the total transmitted bits. It is defined as:

$$BER = \frac{N_e}{N_t} \tag{12}$$

where:

- N_e is the number of erroneous bits received.
- N_t is the total number of transmitted bits.

BER is strongly dependent on the Signal-to-Noise Ratio (SNR), the modulation scheme used, and channel conditions [5, 10].

2.3.1 BER and SNR relationship

In most wireless systems, noise is modeled as Additive White Gaussian Noise (AWGN) with zero mean and variance $N_0/2$. The probability of bit errors is derived using the Q-function, which represents the tail probability of a Gaussian distribution:

$$Q(x) = \frac{1}{\sqrt{2\pi}} \int_x^{\infty} e^{-\frac{t^2}{2}} dt \tag{13}$$

For different modulation schemes, BER is approximated as follows:

BPSK (Binary Phase Shift Keying) BPSK is one of the most robust modulation schemes, offering strong resistance to noise. Its BER is given by:

$$BER_{BPSK} = Q\left(\sqrt{\frac{2E_b}{N_0}}\right) \tag{14}$$

where E_b/N_0 is the energy per bit to noise power spectral density ratio [1].

QPSK (Quadrature Phase Shift Keying) Since QPSK encodes two bits per symbol, its BER is similar to BPSK:

$$BER_{QPSK} = Q\left(\sqrt{\frac{2E_b}{N_0}}\right) \tag{15}$$

M-QAM (Quadrature Amplitude Modulation), higher-order QAM schemes improve spectral efficiency but at the cost of increased BER [9]. The approximate BER for M-QAM is:

$$BER_{M-QAM} \approx \frac{4}{\log_2 M} Q\left(\sqrt{\frac{3}{M-1}\frac{E_b}{N_0}}\right) \tag{16}$$

where M is the modulation order (e.g., 16-QAM, 32-QAM).

2.3.2 BER in fading channels

In practical scenarios, wireless channels experience Rayleigh or Rician fading, causing BER degradation. Under Rayleigh fading, the BER expression for BPSK is:

$$BER_{Rayleigh} = \frac{1}{2}\left(1 - \sqrt{\frac{\gamma}{1+\gamma}}\right) \tag{17}$$

where $\gamma = E_b/N_0$.

2.3.3 Adaptive modulation and BER optimization

Adaptive modulation dynamically selects the most suitable modulation scheme based on real-time SNR conditions, ensuring optimal balance between throughput and error performance:

- High SNR $\rightarrow$ Higher-order modulation (e.g., 32-QAM, 64-QAM) to maximize data rate.
- Low SNR $\rightarrow$ Lower-order modulation (e.g., BPSK, QPSK) to reduce BER.

2.3.4 BER vs. Distance analysis

Since SNR decreases with distance, BER increases correspondingly. The impact of distance on BER can be modeled using:

$$BER(d) \propto Q\left(\sqrt{\frac{2P_t G_t G_r \left(\frac{\lambda}{4\pi d}\right)^2}{N_0}}\right) \tag{18}$$

which shows that at larger distances d, BER worsens due to reduced received power [13, 14].

2.3.5 Impact of SNR on adaptive modulation

SNR significantly influences the choice of modulation schemes in adaptive wireless systems. Higher-order modulation schemes, such as 32-QAM or 64-QAM, offer higher data rates but require a strong SNR to minimize errors. Conversely, low-SNR environments necessitate more robust schemes like BPSK or QPSK. The bit error rate (BER) for a given modulation scheme can be approximated as:

$$BER \approx Q\left(\sqrt{\frac{2E_b}{N_0}}\right) \tag{19}$$

where $Q(\cdot)$ is the Q-function, and E_b/N_0 represents the energy-per-bit to noise ratio [2].

2.3.6 Throughput and SNR relationship

Throughput is a critical performance metric affected by SNR. It is given by:

$$\text{Throughput} = \eta \times B \times \log_2(M) \tag{20}$$

where:

- η is the spectral efficiency,
- B is the available bandwidth,
- M is the modulation order.

Adaptive modulation techniques optimize throughput by selecting the most suitable modulation scheme based on real-time SNR measurements, ensuring an optimal balance between reliability and data rate [1].

3. SNR and path loss model

The received signal power $P_r(d)$ at a distance d can be expressed as:

$$P_r(dB) = P_t + G_t + G_r - PL(d) \tag{21}$$

where:

- P_t is the transmitted power (dBm),
- G_t, G_r are the transmitter and receiver antenna gains (dB),
- $PL(d)$ is the path loss (dB), calculated as:

$$PL(d) = 20\log_{10}(d) + 20\log_{10}(f) - 147.55 \tag{22}$$

where d is the distance in meters and f is the carrier frequency in Hz [18].

4. Computation of SNR

Signal-to-Noise Ratio (SNR) plays a pivotal role in wireless communication systems, serving as a fundamental indicator of signal quality and link reliability [2]. As the distance from the transmitting antenna increases, the received signal power attenuates due to path loss, while the noise power remains relatively constant—leading to a drop in SNR. This degradation directly influences the choice of modulation scheme: higher-order modulation techniques such as 64-QAM or 256-QAM offer greater spectral efficiency but require higher SNR to function reliably. In contrast, robust lower-order schemes like BPSK or QPSK perform better under low-SNR conditions but sacrifice throughput. Consequently, the modulation scheme must adapt dynamically to the SNR, which is itself a function of distance. Moreover, higher-order modulation not only increases susceptibility to noise but also elevates hardware complexity and power consumption, due to more sophisticated signal processing and tighter symbol synchronization requirements. Hence, an intelligent balance between modulation complexity, achievable data rate, and real-time SNR—particularly as influenced by distance—is critical for optimizing the performance and efficiency of modern wireless systems [1].

The SNR is determined as:

$$SNR(dB) = P_r - N_0 \tag{23}$$

where N_0 is the noise power, given by:

$$N_0 = -174 + 10\log_{10}(B) + NF \tag{24}$$

where:

- B is the bandwidth in Hz,
- NF is the receiver noise figure (dB).

5. Simulation and results

5.1 Analysis of signal modulation and distance

In this section, we present an analytical framework for evaluating the performance of various modulation schemes over different distances using MATLAB simulations. The analysis primarily focuses on key performance metrics such as the Signal-to-Noise Ratio (SNR), Bit Error Rate (BER), and achievable throughput. Additionally, a novel parameter, the extitModulation Adaptability Score (MAS), is introduced to assess the suitability of modulation schemes under varying channel conditions.

5.2 Path loss and SNR calculation

In wireless communication systems, path loss is a fundamental factor that directly impacts the Signal-to-Noise Ratio (SNR). Path loss refers to the attenuation of signal power as it propagates through space, primarily due to the spreading of the wavefront and environmental factors such as absorption, diffraction, and reflection. As described by the Free-Space Path Loss (FSPL) model, the received signal power diminishes proportionally to the square of the distance between the transmitter and receiver, leading to a logarithmic decline in SNR. Mathematically, as the distance d increases, the path loss increases by $20\log_{10}(d)$, and since SNR is defined as the ratio of received signal power to noise power, a higher path loss translates to a lower SNR. This reduction in SNR adversely affects the system's ability to support high-order modulation schemes, increasing the Bit Error Rate (BER) and reducing the achievable data rate. Therefore, understanding the path loss and its influence on SNR is crucial for designing adaptive communication strategies that maintain performance and reliability over varying transmission distances.

The Free-Space Path Loss (FSPL) model is used to compute the received signal power over increasing distances, given by:

$$FSPL(dB) = 20\log_{10}(d) + 20\log_{10}(f) + 20\log_{10}\left(\frac{4\pi}{c}\right) \tag{25}$$

where d is the distance in meters, f is the carrier frequency in Hz, and c is the speed of light in m/s. The received power P_r is derived from the transmitted power P_t and FSPL:

$$P_r(dBm) = P_t(dBm) - FSPL(dB) \tag{26}$$

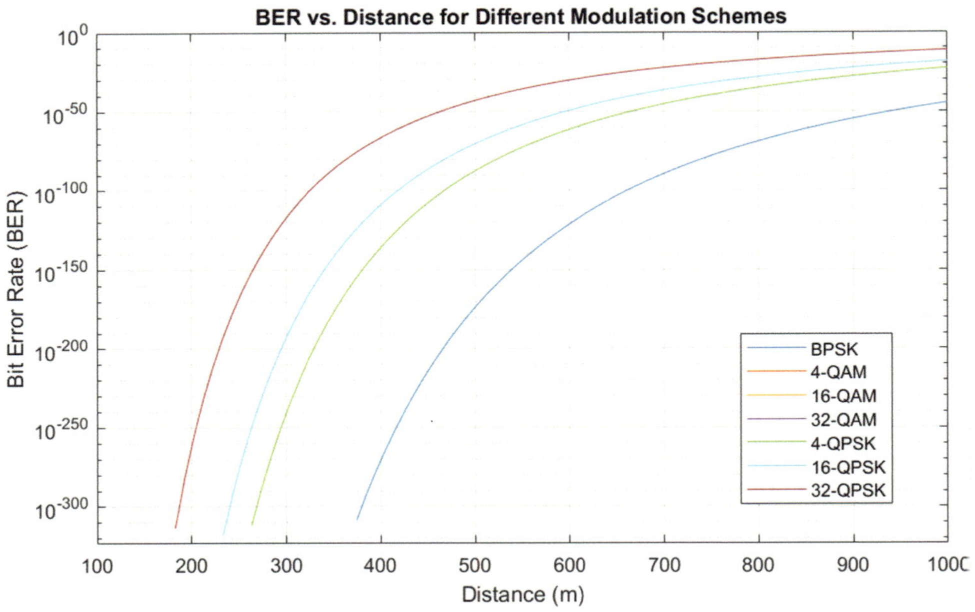

Figure 4.
BER versus distance: comparison of the modulation schemes.

The SNR is calculated using the noise power, considering a given noise figure and thermal noise power.

Figure 4. Illustrates the BER variation with distance for different modulation schemes. The key observations from this analysis are:

- Higher-order modulation schemes (e.g., 32-QAM, 32-QPSK) exhibit higher BER at a given distance due to their lower noise tolerance.
- Lower-order schemes (e.g., BPSK, 4-QAM) perform better at longer distances as they require lower SNR for reliable transmission.
- The BER increases exponentially with distance, emphasizing the necessity of adaptive modulation to optimize performance.

6. BER computation for different modulation schemes

The BER is computed for various modulation schemes using their respective theoretical formulations:

- **BPSK:** $BER = \frac{1}{2}\mathrm{erfc}\left(\sqrt{SNR}\right)$
- **QPSK:** $BER = \frac{1}{2}\mathrm{erfc}\left(\sqrt{SNR/2}\right)$
- **M-QAM:** $BER = \frac{4}{\log_2 M}\left(1 - \frac{1}{\sqrt{M}}\right)\mathrm{erfc}\left(\sqrt{\frac{3SNR}{2(M-1)}}\right)$

6.1 Throughput analysis

The achievable data rate R is given by:

$$R = B \cdot \log_2(M) \cdot (1 - BER) \tag{27}$$

where B is the bandwidth and M represents the modulation order.

6.2 Modulation adaptability score (mas)

A novel parameter, MAS, is defined to evaluate the efficiency of modulation schemes over varying distances:

$$MAS = \frac{R}{BER + \epsilon} \tag{28}$$

where ε is a small value to prevent division by zero. A higher MAS value indicates better adaptability to changing distance conditions [15, 16].

6.3 Reflection on results

The MATLAB code simulates these metrics and plots BER, Throughput, and MAS against distance. The results provide insight into the effectiveness of adaptive modulation over fixed schemes, supporting the case for distance-based modulation adaptation in wireless systems.

Using MATLAB, the SNR was computed over a range of distances (10 m to 1,000 m). The resulting plot is shown in **Figure 5**.

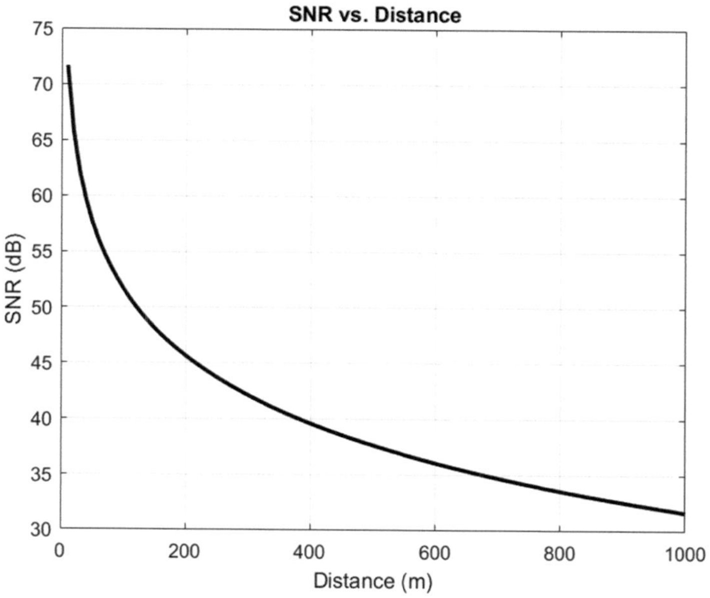

Figure 5.
SNR versus distance from antenna.

The key observations include:

- **At short distances**, the SNR is high, allowing the use of high-order modulation schemes such as 32-QAM.
- **As distance increases**, the SNR decreases logarithmically, necessitating a transition to lower-order schemes such as QPSK or BPSK.
- **Beyond a certain distance**, only robust modulation schemes with lower data rates can be supported.

7. Conclusion

This chapter proposed a distance-aware adaptive modulation strategy aimed at improving wireless communication performance under varying channel conditions. By dynamically selecting the most suitable modulation scheme based on the estimated transmitter-receiver distance and corresponding signal quality, the system demonstrates improved adaptability and spectral efficiency. Simulation results showed that the proposed technique yields tangible improvements in Bit Error Rate (BER) performance across multiple distance ranges and propagation environments.

A key contribution of this work lies in its practical approach—leveraging fundamental parameters such as path loss and SNR—to drive intelligent modulation adaptation without requiring additional hardware or complex feedback loops. This makes the solution particularly attractive for resource-constrained environments such as IoT deployments and mobile terminals operating under diverse link conditions.

Nonetheless, several limitations warrant further exploration. The current study relies solely on MATLAB-based simulations and assumes idealized propagation models. Real-world deployment factors such as multipath fading, shadowing, hardware non-linearities, and dynamic user mobility are not fully captured. Additionally, while the Modulation Adaptability Score (MAS) provides a useful index for modulation decision-making, it may benefit from calibration or weighting schemes tailored to application-specific requirements (e.g., latency-sensitive versus throughput-sensitive scenarios).

Future Work. To extend this research, several promising directions can be explored:

- **Experimental Validation:** Implement and test the proposed scheme on a software-defined radio (SDR) platform (e.g., USRP or NI PXI) in real indoor and outdoor environments to validate its robustness under practical wireless conditions.
- **Machine Learning Integration:** Incorporate reinforcement learning or neural networks to predict optimal modulation schemes based on historical SNR and path loss trends, allowing for faster adaptation and improved resilience in rapidly changing environments.

- **Multi-user and MIMO Extensions:** Extend the adaptive strategy to multi-user and MIMO systems, analyzing its impact on spatial diversity and system-wide resource allocation.

- **Cross-layer Optimization:** Investigate how the modulation adaptation interacts with higher-layer protocols (e.g., MAC and transport) and optimize system performance from a cross-layer perspective.

- **Threshold Optimization:** Develop environment-aware or learning-based methods for dynamically adjusting SNR and path loss thresholds based on context, such as device mobility, user density, or interference level.

In summary, while the proposed distance-based adaptive modulation framework offers a practical and efficient means of improving wireless link performance, there is ample opportunity to expand, generalize, and experimentally validate its capabilities in future work.

Acknowledgment

This publication has emanated from research conducted with the financial support of Taighde Éireann – Research Ireland under Grant number 13/RC/2077_P2 at CONNECT: the Research Ireland Centre for Future Network.

Conflict of Interest

The author declares no conflict of interest.

Thanks

Heartfelt thanks to my mother, who supports my growth from afar with love and strength. To my wonderful children; thank you for your patience and joy. To my husband, Dr. Pooria Varahram, whose unwavering support means everything. And to my late father, who always called me “دختر شجاع من” which means “my brave girl” and gave me the confidence to believe in myself every single day.

Nomenclature

d	Transmitter-receiver distance (m)
f	Carrier frequency (Hz)
c	Speed of light, 3×10^8 m/s
PL	Path loss (dB)
PL_0	Reference path loss at distance d_0 (dB)
n	Path loss exponent
P_t	Transmit power (dBm)
P_r	Received power (dBm)

G_t	Transmitter antenna gain (dB)
G_r	Receiver antenna gain (dB)
N_0	Noise power (dBm)
B	Channel bandwidth (Hz)
SNR	Signal-to-Noise Ratio (linear)
SNR_{dB}	Signal-to-Noise Ratio (dB)
R	Achievable data rate (bps)
M	Modulation order
η	Spectral efficiency
BER	Bit Error Rate
E_b/N_0	Energy-per-bit to noise power spectral density ratio
$Q(\cdot)$	Q-function used in probability of error calculations
MAS	Modulation Adaptability Score
λ	Wavelength of the carrier signal (m)

Abbreviations

BER	Bit Error Rate
BPSK	Binary Phase Shift Keying
QPSK	Quadrature Phase Shift Keying
QAM	Quadrature Amplitude Modulation
SNR	Signal-to-Noise Ratio
FSPL	Free Space Path Loss
AMC	Adaptive Modulation and Coding
MAS	Modulation Adaptability Score
MIMO	Multiple Input Multiple Output
AWGN	Additive White Gaussian Noise
IoT	Internet of Things
5G	Fifth Generation Wireless Network

Author details

Somayeh Mohammady
Technological University Dublin (TUDublin), Dublin, Ireland

*Address all correspondence to: somayeh.mohammady@tudublin.ie

References

[1] Tse D, Viswanath P. Fundamentals of Wireless Communication University of California, Berkeley, Pramod Viswanath, University of Illinois, Urbana-Champaign: Cambridge University Press; 2005

[2] Goldsmith A. Wireless Communications. Stanford University, California: Cambridge University Press; 2005

[3] Shannon CE. A mathematical theory of communication. Bell System Technical Journal. 1948;**27**(3):379–423

[4] Rahhal J, Rahhal S. Cluster modulation: A generalized modulation scheme leading to NOMA or adaptive modulation. Journal of Computer and Communications. 2024;**12**:12–22. DOI: 10.4236/jcc.2024.127002

[5] Proakis JG, Salehi M. Digital Communications. 5th ed. University of California, San Diego: McGraw-Hill; 2007

[6] Haykin S, Moher M. Modern Wireless Communications. USA: Pearson Education; 2001

[7] Randall JR, Arisheh AA, Merlo JM, Nanzer JA. "A Dynamic Array Using Spatial Amplitude Modulation With an Asymmetric Wilkinson Power Divider for Secure Wireless Applications". In: IEEE Antennas and Wireless Propagation Letters. 2023;**22**(12):3107–3111. DOI: 10.1109/LAWP.2023.3310911

[8] June Chul Roh and BD Rao, "Adaptive modulation for multiple antenna channels," Conference Record of the Thirty-Sixth Asilomar Conference on Signals, Systems and Computers, 2002., vol. **1**. Pacific Grove, CA, USA: 2002. p. 526–530. DOI: 10.1109/ACSSC.2002.1197237

[9] Rappaport TS, Xing Y, MacCartney GR, Molisch AF, Mellios E, Zhang J. Overview of millimeter wave communications for fifth-generation (5G) wireless networks with a focus on propagation models. IEEE Transactions on Antennas and Propagation. 2017;**65**(12):6213–6230. DOI: 10.1109/TAP.2017.2734243

[10] Molisch AF. Wireless Communications. University of Southern California, USA: 2nd ed. Wiley-IEEE Press; 2011

[11] Rappaport TS, Xing Y, MacCartney GR, et al. Overview of Millimeter Wave Communications for Fifth-Generation (5G) Wireless Networks with a Focus on Propagation Models. IEEE Transactions on Antennas and Propagation 2017;**65**:6213–6230

[12] Centenaro M, Vangelista L, Zanella A, Zorzi M. Long-Range Communications in Unlicensed Bands: The Rising Stars in the IoT and Smart City Scenarios. IEEE Wireless Communications. 2016;**23**(5):60–67

[13] Luo S, Yang P, Che YL, Wu K, Peng Y, Li S. "Generalized Space Domain Index Modulation for mmWave Distributed Antenna Systems". IEEE Transactions on Vehicular Technology. 2020;**69**(11):14067–14071. DOI: 10.1109/TVT.2020.3030628

[14] Centenaro M, Vangelista L, Zanella A, Zorzi M. Long-range communications in unlicensed bands: The rising stars in the IoT and smart city

scenarios. IEEE Wireless Communications. 2016;**23**(5):60–67. DOI: 10.1109/MWC.2016.7721743

[15] Sun S, Rappaport TS, Thomas TA, Ghosh A, Nguyen HC, Kovacs IZ, et al. Investigation of prediction accuracy, sensitivity, and parameter stability of large-scale propagation path loss models for 5G wireless communications. IEEE Transactions on Vehicular Technology. 2016;**65**(5):2843–2860. DOI: 10.1109/TVT.2016.2543139

[16] Goldsmith AJ, Chua SG. Variable-rate variable-power MQAM for fading channels. IEEE Transactions on Communications. 1997;**45**(10):1218–1230. DOI: 10.1109/26.634685

[17] Bai T, Heath RW Jr. Coverage and rate analysis for millimeter-wave cellular networks. IEEE Transactions on Wireless Communications. 2015;**14**(2):1100–1114. DOI: 10.1109/TWC.2014.2364267

[18] Bazzi A, et al. "Upper mid-band spectrum characteristics and path loss insights toward 6G deployment," IEEE Communications Magazine, 2025

[19] Liu Q, Zhou S, Giannakis GB. Cross-layer combining of adaptive modulation and coding with truncated ARQ over wireless links. IEEE Transactions on Wireless Communications. 2004;**3**(5):1746–1755. DOI: 10.1109/TWC.2004.833474

[20] Singh S, Ziliotto F, Madhow U, Belding E, Rodwell M. "Blockage and directivity in 60 GHz wireless personal area networks: from cross-layer model to multihop MAC design". in IEEE Journal on Selected Areas in Communications. 2009;**27**(8):1400–1413. DOI: 10.1109/JSAC.2009.091010